ASE Test Preparation

Automotive Technician Certification Series

Manual Drive Trains and Axles (A3)
5th Edition

DELMAR
CENGAGE Learning

Australia • Brazil • Japan • Korea • Mexico • Singapore • Spain • United Kingdom • United States

DELMAR
CENGAGE Learning™

ASE Test Preparation: Automotive Technician Certification Series, Manual Drive Trains and Axles (A3), 5th Edition

Vice President, Technology and Trades Professional Business Unit: Gregory L. Clayton

Director, Professional Transportation Industry Training Solutions: Kristen L. Davis

Product Manager: Katie McGuire

Editorial Assistant: Danielle Filippone

Director of Marketing: Beth A. Lutz

Marketing Manager: Jennifer Barbic

Senior Production Director: Wendy Troeger

Content Project Manager: PreMediaGlobal

Senior Art Director: Benj Gleeksman

Section Opener Image: Image Copyright Creations, 2012. Used under License from Shutterstock.com

For product information and technology assistance, contact us at
Cengage Learning Customer & Sales Support, 1-800-354-9706
For permission to use material from this text or product, submit all requests online at **www.cengage.com/permissions**
Further permissions questions can be e-mailed to
permissionrequest@cengage.com

ISBN-13: 978-1-111-12705-3

ISBN-10: 1-111-12705-0

Delmar Cengage Learning
Executive Woods
5 Maxwell Drive
Clifton Park, NY 12065-2919
USA

Cengage Learning is a leading provider of customized learning solutions with office locations around the globe, including Singapore, the United Kingdom, Australia, Mexico, Brazil, and Japan. Locate your local office at **www.cengage.com/global**

Cengage Learning products are represented in Canada by Nelson Education, Ltd.

For more information on transportation titles available from Delmar Cengage Learning, please visit our website at **www.trainingbay.cengage.com**

Visit our corporate website at **www.cengage.com**

Notice to the Reader

Printed in the USA
6 7 8 9 10   27 26 25 24 23

Table of Contents

Delmar, a part of Cengage Learning, is very pleased that you have chosen to use our ASE Test Preparation Guide to help prepare yourself for the Manual Drive Trains and Axles (A3) ASE certification examination. This guide is designed to help prepare you for your actual exam by providing you with an overview and introduction of the testing process, introducing you to the task list for the Manual Drive Trains and Axles (A3) certification exam, giving you an understanding of what knowledge and skills you are expected to have in order to successfully perform the duties associated with each task area, and providing you with several preparation exams designed to emulate the live exam content in hopes of assessing your overall exam readiness.

If you have a basic working knowledge of the discipline you are testing for, you will find this book will be an excellent guide, helping you to understand the "must know" items needed to successfully pass the ASE certification exam. This manual is not a textbook. Its objective is to prepare the individual who has the existing requisite experience and knowledge to attempt the challenge of the ASE certification process. This guide cannot replace the hands-on experience and theoretical knowledge required by ASE to master the vehicle repair technology associated with this exam. If you are unable to understand more than a few of the preparation questions and their corresponding explanations in this book, it could be that you require either more shop-floor experience or further study.

This book begins by providing an overview of, and introduction to, the testing process. This section outlines what we recommend you do to prepare, what to expect on the actual test day, and overall methodologies for your success. This section is followed by a detailed overview of the ASE task list to include explanations of the knowledge and skills you must possess to successfully answer questions related to each particular task. After the task list, we provide six sample preparation exams for you to use as a means of evaluating areas of understanding, as well as areas requiring improvement in order to successfully pass the ASE exam. Delmar is the first and only test preparation organization to provide so many unique preparation exams. We enhanced our guides to include this support as a means of providing you with the best preparation product available. Section 6 of this guide includes the answer keys for each preparation exam, along with the answer explanations for each question. Each answer explanation also contains a reference back to the related task or tasks that it assesses. This will provide you with a quick and easy method for referring back to the task list whenever needed. The last section of this book contains blank answer sheet forms you can use as you attempt each preparation exam, along with a glossary of terms.

OUR COMMITMENT TO EXCELLENCE

Thank you for choosing Delmar, Cengage Learning for your ASE test preparation needs. All of the writers, editors, and Delmar staff have worked very hard to make this test preparation guide second to none. We feel confident that you will find this guide easy to use and extremely beneficial as you prepare for your actual ASE exam.

Delmar, Cengage Learning has sought out the best subject-matter experts in the country to help with the development of *ASE Test Preparation: Automotive Technician Certification Series, Manual Drive Trains and Axles (A3), 5th Edition*. Preparation questions are authored and then reviewed by a group of certified, subject-matter experts to ensure the highest level of quality and validity to our product.

If you have any questions concerning this guide or any guide in this series, please visit us on the web at **http://www.trainingbay.cengage.com**.

For online test preparation solutions for ASE certifications, please visit us on the web at **http://www.techniciantestprep.com** to learn more.

ABOUT THE AUTHOR

Doug Poteet has been around all types of vehicles his entire life. He was raised on a small farm in central Kentucky, where his father, Gordon, got him interested in fixing cars, trucks, and tractors at a young age. Doug earned his diploma from Nashville Auto/Diesel College at age 17 and went on to become a heavy truck technician; after several years as a truck technician, Doug became a line technician at a new car dealership. For the next 15 years, he specialized in driveability. In 1994, he joined the faculty at Elizabethtown Community and Technical College, where he currently serves as Associate Professor for the Automotive and Diesel programs. He earned an Associate of Science degree in Vocational/Technical Education from Western Kentucky University in Bowling Green. Doug holds ASE certifications in Master Automotive Technician, Master Medium/Heavy Truck Technician, School Bus Technician, Advanced Engine Performance, and Parts Specialist. In his free time, Doug enjoys hunting, climbing hills in his rail buggy with his wife, Sandy, and spending time with his grown sons, Kirk and Wesley.

ABOUT THE SERIES ADVISOR

Mike Swaim has been an Automotive Technology Instructor at North Idaho College, Coeur d'Alene, Idaho since 1978. He is an Automotive Service Excellence (ASE) Certified Master Technician since 1974 and holds a Lifetime Certification from the Mobile Air Conditioning Society. He served as Series Advisor to all nine of the 2011 Automotive Technician/Light Truck Technician Certification Tests (A Series) of Delmar, Cengage Learning ASE Test Preparation titles, and is the author of *ASE Test Preparation: Automotive Technician Certification Series, Undercar Specialist Designation (X1), 5th Edition*.

The History and Purpose of ASE

ASE began as the National Institute for Automotive Service Excellence (NIASE). It was founded as a non-profit, independent entity in 1972 by a group of industry leaders with the single goal of providing a means for consumers to distinguish between incompetent and competent technicians. It accomplishes this goal through the testing and certification of repair and service professionals. Though it is still known as the National Institute for Automotive Service Excellence, it is now called "ASE" for short.

Today, ASE offers more than 40 certification exams in automotive, medium/heavy duty trucks, collision repair and refinish, school bus, transit bus, parts specialist, automobile service consultant and other industry-related areas. At this time, there are more than 385,000 professionals nationwide with current ASE certifications. These professionals are employed by new car and truck dealerships, independent repair facilities, fleets, service stations, franchised service facilities and more.

ASE's certification exams are industry-driven and cover practically every on-highway vehicle service segment. The exams are designed to stress the knowledge of job-related skills. Certification consists of passing at least one exam and documenting two years of relevant work experience. To maintain certification, those with ASE credentials must be re-tested every five years.

While ASE certifications are a targeted means of acknowledging the skills and abilities of an individual technician, ASE also has a program designed to provide recognition for highly qualified repair, support and parts businesses. The Blue Seal of Excellence Recognition Program allows businesses to showcase their technicians and their commitment to excellence. One of the requirements of becoming Blue Seal recognized is that the facility must have a minimum of 75 percent of their technicians ASE certified. Additional criteria apply, and program details can be found on the ASE website.

ASE recognized that educational programs serving the service and repair industry also needed a way to be recognized as having the faculty, facilities and equipment to provide a quality education to students wanting to become service professionals. Through the combined efforts of ASE, industry and education leaders, the non-profit organization entitled the National Automotive Technicians Education Foundation (NATEF) was created in 1983 to evaluate and recognize academic programs. Today more than 2,000 educational programs are NATEF certified.

For additional information about ASE, NATEF or any of their programs, the following contact information can be used:

National Institute for Automotive Service Excellence (ASE)

101 Blue Seal Drive S.E.

Suite 101

Leesburg, VA 20175

Telephone: 703-669-6600

Fax: 703-669-6123

Website: **www.ase.com**

Overview and Introduction

Participating in the National Institute for Automotive Service Excellence (ASE) voluntary certification program provides you with the opportunity to demonstrate you are a qualified and skilled professional technician who has the "know-how" required to successfully work on today's modern vehicles.

EXAM ADMINISTRATION

Note: After November 2011, ASE will no longer offer paper and pencil certification exams. There will be no Winter testing window in 2012, and ASE will offer and support CBT testing exclusively starting in April 2012.

ASE provides computer-based testing (CBT) exams, which are administered at test centers across the nation. It is recommended that you go to the ASE website at *http://www.ase.com* and review the conditions and requirements for this type of exam. There is also an exam demonstration page that allows you to personally experience how this type of exam operates before you register.

CBT exams are available four times annually, for two-month windows, with a month of no testing in between each testing window:

- January/February – Winter testing window
- April/May – Spring testing window
- July/August – Summer testing window
- October/November – Fall testing window

Please note, testing windows and timing may change. It is recommended you go to the ASE website at *http://www.ase.com* and review the latest testing schedules.

UNDERSTANDING TEST QUESTION BASICS

ASE exam questions are written by service industry experts. Each question on an exam is created during an ASE-hosted "item-writing" workshop. During these workshops, expert service representatives from manufacturers (domestic and import), aftermarket parts and equipment manufacturers, working technicians, and technical educators gather to share ideas and convert them into actual exam questions. Each exam question written by these experts must then survive review by all members of the group. The questions are designed to address the practical application of repair and diagnosis knowledge and skills practiced by technicians in their day-to-day work.

After the item-writing workshop, all questions are pre-tested and quality-checked on a national sample of technicians. Those questions that meet ASE standards of quality and accuracy are included in the scored sections of the exams; the "rejects" are sent back to the drawing board or discarded altogether.

Depending on the topic of the certification exam, you will be asked between 40 and 80 multiple-choice questions. You can determine the approximate number of questions you can expect to be asked during the Manual Drive Trains and Axles (A3) certification exam by reviewing the task list in Section 4 of this book. The five-year recertification exam will cover this same content; however, the number of questions for each content area of the recertification exam will be reduced by approximately one-half.

> *Note:* Exams may contain questions that are included for statistical research purposes only. Your answers to these questions will not affect your score, but since you do not know which ones they are, you should answer all questions in the exam.

Using multiple criteria, including cross-sections by age, race, and other background information, ASE is able to guarantee that exam questions do not include bias for or against any particular group. A question that shows bias toward any particular group is discarded.

TEST-TAKING STRATEGIES

Before beginning your exam, quickly look over the exam to determine the total number of questions that you will need to answer. Having this knowledge will help you manage your time throughout the exam to ensure you have enough available to answer all of the questions presented. Read through each question completely before marking your answer. Answer the questions in the order they appear on the exam. Leave the questions blank that you are not sure of and move on to the next question. You can return to those unanswered questions after you have finished the others. These questions may actually be easier to answer at a later time, once your mind has had additional time to consider them on a subconscious level. In addition, you might find information in other questions that will help you recall the answers to some of them.

Multiple-choice exams are sometimes challenging because there are often several choices that may seem possible, or partially correct, and therefore it may be difficult to decide on the most appropriate answer choice. The best strategy, in this case, is to first determine the correct answer before looking at the answer options. If you see the answer you decided on, you should still be careful to examine the other answer options to make sure that none seems more correct than yours. If you do not know or are not sure of the answer, read each option very carefully and try to eliminate those options that you know are incorrect. That way, you can often arrive at the correct choice through a process of elimination.

If you have gone through the entire exam, and you still do not know the answer to some of the questions, *then guess*. Yes, guess. You then have at least a 25 percent chance of being correct. While your score is based on the number of questions answered correctly, any question left blank, or unanswered, is automatically scored as incorrect.

There is a lot of "folk" wisdom on the subject of test taking that you may hear about as you prepare for your ASE exam. For example, there are those who would advise you to avoid response options that use certain words such as *all, none, always, never, must,* and *only,* to name a few. This, they claim, is because nothing in life is exclusive. They would advise you to choose response options that use words that allow for some exception, such as *sometimes, frequently, rarely, often, usually, seldom,* and *normally.* They would also advise you to avoid the first and last option (A or D) because exam writers, they feel, are more comfortable if they put the correct answer in the middle (B or C) of the choices. Another recommendation often offered is to select the option that is either shorter or longer than the other three choices because it is more likely to be correct. Some would advise you to never change an answer since your first intuition is usually correct. Another area of "folk" wisdom focuses specifically on any repetitive patterns created by your question responses (e.g., A, B, C, A, B, C, A, B, C).

Many individuals may say that there are actual grains of truth in this "folk" wisdom, and whereas with some exams, this may prove true, it is not relevant in regard to the ASE certification exams.

ASE validates all exam questions and test forms through a national sample of technicians, and only those questions and test forms that meet ASE standards of quality and accuracy are included in the scored sections of the exams. Any biased questions or patterns are discarded altogether, and therefore, it is highly unlikely you will experience any of this "folk" wisdom on an actual ASE exam.

PREPARING FOR THE EXAM

Delmar, Cengage Learning wants to make sure we are providing you with the most thorough preparation guide possible. To demonstrate this, we have included hundreds of preparation questions in this guide. These questions are designed to provide as many opportunities as possible to prepare you to successfully pass your ASE exam. The preparation approach we recommend and outline in this book is designed to help you build confidence in demonstrating what task area content you already know well while also outlining what areas you should review in more detail prior to the actual exam.

We recommend that your first step in the preparation process should be to thoroughly review Section 3 of this book. This section contains a description and explanation of the type of questions you'll find on an ASE exam.

Once you understand how the questions will be presented, we then recommend that you thoroughly review Section 4 of this book. This section contains information that will help you establish an understanding of what the exam will be evaluating, and specifically, how many questions to expect in each specific task area.

As your third preparatory step, we recommend you complete your first preparation exam, located in Section 5 of this book. Answer one question at a time. After you answer each question, review the answer and question explanation information, located in Section 6. This section will provide you with instant response feedback, allowing you to gauge your progress, one question at a time, throughout this first preparation exam. If after reading the question explanation you do not feel you understand the reasoning for the correct answer, go back and review the task list overview (Section 4) for the task that is related to that question. Included with each question explanation is a clear identifier of the task area that is being assessed (e.g., Task A.1). If at that point you still do not feel you have a solid understanding of the material, identify a good source of information on the topic, such as an educational course, textbook or other related source of topical learning, and do some additional studying.

After you have completed your first preparation exam and have reviewed your answers, you are ready to complete your next preparation exam. A total of six practice exams are available in Section 5 of this book. For your second preparation exam, we recommend that you answer the questions as if you were taking the actual exam. Do not use any reference material or allow any interruptions in order to get a feel for how you will do on the actual exam. Once you have answered all of the questions, grade your results using the answer key in Section 6. For every question that you gave an incorrect answer to, study the explanations to the answers and/or the overview of the related task areas. Try to determine the root cause for missing the question. The easiest thing to correct is learning the correct technical content. The hardest things to correct are behaviors that lead you to an incorrect conclusion. If you knew the information but still got the question incorrect, there is likely a test-taking behavior that will need to be corrected. An example of this would be reading too quickly and skipping over words that affect your reasoning. If you can identify what you did that caused you to answer the question incorrectly, you can eliminate that cause and improve your score.

Here are some basic guidelines to follow while preparing for the exam:

- Focus your studies on those areas you are weak in.
- Be honest with yourself when determining if you understand something.
- Study often but for short periods of time.
- Remove yourself from all distractions when studying.
- Keep in mind that the goal of studying is not just to pass the exam; the real goal is to learn.
- Prepare physically by getting a good night's rest before the exam, and eat meals that provide energy but do not cause discomfort.
- Arrive early to the exam site to avoid long waits as test candidates check in.
- Use all of the time available for your exams. If you finish early, spend the remaining time reviewing your answers.
- Do not leave any questions unanswered. If absolutely necessary, guess. All unanswered questions are automatically scored as incorrect.

Here are some items you will need to bring with you to the exam site:

- A valid government or school-issued photo ID
- Your test center admissions ticket
- A watch (not all test sites have clocks)

> *Note:* Books, calculators, and other reference materials are not allowed in the exam room. The exceptions to this list are English-Foreign dictionaries or glossaries. All items will be inspected before and after testing.

WHAT TO EXPECT DURING THE EXAM

When taking a CBT exam, as soon as you are seated in the testing center, you will be given a brief tutorial to acquaint you with the computer-delivered test prior to taking your certification exam(s). The CBT exams allow you to select only one answer per question. You can also change your answers as many times as you like. When you select a second answer choice, the CBT will automatically unselect your first answer choice. If you want to skip a question to return to later, you can utilize the "flag" feature, which will allow you to quickly identify and review questions whenever you are ready. Prior to completing your exam, you will also be provided with an opportunity to review your answers and address any unanswered questions.

TESTING TIME

CBT Exams

Each individual ASE CBT exam has a fixed time limit. Individual exam times will vary based upon exam area, and will range anywhere from a half hour to two hours. You will also be given an additional 30 minutes beyond what is allotted to complete your exams to ensure you have adequate time to perform all necessary check-in procedures, complete a brief CBT tutorial, and potentially complete a post-test survey.

You can register for and take multiple CBT exams during one testing appointment. The maximum time allotment for a CBT appointment is four and a half hours. If you happen to register for so many exams that you will require more time than this, your exams will be scheduled into multiple appointments. This could mean that you have testing on both the morning and afternoon of the same day, or they could be scheduled on different days, depending on your personal preference and the test center's schedule.

It is important to understand that if you arrive late for your CBT test appointment, you will not be able to make up any missed time. You will only have the scheduled amount of time remaining in your appointment to complete your exam(s).

Also, while most people finish their CBT exams within the time allowed, others might feel rushed or not be able to finish the test, due to the implied stress of a specific, individual time limit allotment. Before you register for the CBT exams, you should review the number of exam questions that will be asked along with the amount of time allotted for that exam to determine whether you feel comfortable with the designated time limitation or not.

As an overall time management recommendation, you should monitor your progress and set a time limit you will follow with regard to how much time you will spend on each individual exam question. This should be based on the total number of questions you will be answering.

Also, it is very important to note that if for any reason you wish to leave the testing room during an exam, you must first ask permission. If you happen to finish your exam(s) early and wish to leave the testing site before your designated session appointment is completed, you are permitted to do so only during specified dismissal periods.

UNDERSTANDING HOW YOUR EXAM IS SCORED

You can gain a better perspective about the ASE certification exams if you understand how they are scored. ASE exams are scored by an independent organization having no vested interest in ASE or in the automotive industry. With CBT exams, you will receive your exam scores immediately.

Each question carries the same weight as any other question. For example, if there are 50 questions, each is worth 2 percent of the total score.

Your exam results can tell you:

- Where your knowledge equals or exceeds that needed for competent performance, or
- Where you might need more preparation.

Your ASE exam score report is divided into content "task" areas; it will show the number of questions in each content area and how many of your answers were correct. These numbers provide information about your performance in each area of the exam. However, because there may be a different number of questions in each content area of the exam, a high percentage of correct answers in an area with few questions may not offset a low percentage in an area with many questions.

It should be noted that one does not "fail" an ASE exam. The technician who does not pass is simply told "More Preparation Needed." Though large differences in percentages may indicate problem areas, it is important to consider how many questions were asked in each area. Since each exam evaluates all phases of the work involved in a service specialty, you should be prepared in each area. A low score in one area could keep you from passing an entire exam. If you do not pass the exam, you may take it again at any time it is scheduled to be administered.

There is no such thing as average. You cannot determine your overall exam score by adding the percentages given for each task area and dividing by the number of areas. It does not work that way because there generally are not the same number of questions in each task area. A task area with 20 questions, for example, counts more toward your total score than a task area with 10 questions.

Your exam report should give you a good picture of your results and a better understanding of your strengths and areas needing improvement for each task area.

Types of Questions on an ASE Exam

Understanding not only what content areas will be assessed during your exam, but how you can expect exam questions to be presented will enable you to gain the confidence you need to successfully pass an ASE certification exam. The following examples will help you recognize the types of question styles used in ASE exams and assist you in avoiding common errors when answering them.

Most initial certification tests are made up of between 40 and 80 multiple-choice questions. The five-year recertification exams will cover the same content as the initial exam; however, the actual number of questions for each content area will be reduced by approximately one-half. Refer to Section 4 of this book for specific details regarding the number of questions to expect during the initial Manual Drive Trains and Axles (A3) certification exam.

Multiple-choice questions are an efficient way to test knowledge. To correctly answer them, you must consider each answer choice as a possibility, and then choose the answer choice that *best* addresses the question. To do this, read each word of the question carefully. Do not assume you know what the question is asking until you have finished reading the entire question.

About 10 percent of the questions on an actual ASE exam will reference an illustration. These drawings contain the information needed to correctly answer the question. The illustration should be studied carefully before attempting to answer the question. When the illustration is showing a system in detail, look over the system and try to figure out how the system works before you look at the question and the possible answers. This approach will ensure that you do not answer the question based upon false assumptions or partial data, but instead have reviewed the entire scenario being presented.

MULTIPLE-CHOICE/DIRECT QUESTIONS

The most common type of question used on an ASE exam is the direct multiple-choice style question. This type of question contains an introductory statement, called a stem, followed by four options: three incorrect answers, called distracters, and one correct answer, the key.

When the questions are written, the point is to make the distracters plausible to draw an inexperienced technician to inadvertently select one of them. This type of question gives a clear indication of the technician's knowledge.

Here is an example of a direct style question:

1. Which of the following would be used to measure transaxle input shaft end-play?

    A. Dial indicator
    B. Plastigauge®
    C. Outside micrometer
    D. Dial caliper

Answer A is correct. A dial indicator is used to measure end-play or runout.

Answer B is incorrect. Plastigauge® is the most common method of measuring crankshaft main bearing clearance on an engine.

Answer C is incorrect. An outside micrometer is used to measure the outside dimension of a part.

Answer D is incorrect. A dial caliper is used to measure outside depth or inside dimensions.

COMPLETION QUESTIONS

A completion question is similar to the direct question except the statement may be completed by any one of the four options to form a complete sentence.

Here is an example of a completion question:

TASK A.9

1. A vehicle with an electric shift transfer case has a shift motor diagnostic trouble code (DTC). The technician wiggles the wiring harness while observing the shift motor voltage with the scan tool. The shift motor voltage changes. The most likely cause of the DTC:

    A. Faulty scan tool.
    B. Faulty transfer case shift control module.
    C. Faulty shift motor wiring.
    D. Faulty transfer case shift control module power supply.

TASKS B.15 & F.11

Answer A is incorrect. If the voltage value changed while moving the wiring harness, there is no reason to believe the scan tool is faulty.

Answer B is incorrect. If the voltage value changed while moving the wiring harness, there is no reason to believe the transfer case shift control module is faulty.

Answer C is correct. If the voltage value changed while moving the wiring harness, the most likely cause is the wiring harness.

Answer D is incorrect. If all other items associated with the transfer case shift control module are normal and the voltage value changed while moving the wiring harness, the most likely cause is the wiring harness.

TECHNICIAN A, TECHNICIAN B QUESTIONS

This type of question is usually associated with an ASE exam. It is, in fact, two true-false statements grouped together, such as: "Technician A says…" and "Technician B says…", followed by "Who is correct?"

In this type of question, you must determine whether either, both, or neither of the statements are correct. To answer this type of question correctly, you must carefully read each technician's statement and judge it on its own merit.

Sometimes this type of question begins with a statement about some analysis or repair procedure. This statement provides the setup or background information required to understand the conditions about which Technician A and Technician B are talking, followed by two statements about the cause of the concern, proper inspection, identification, or repair choices.

Analyzing this type of question is a little easier than the other types because there are only two ideas to consider, although there are still four choices for an answer.

Again, Technician A, Technician B questions are really double true-or-false questions. The best way to analyze this type of question is to consider each technician's statement separately. Ask yourself, "Is A true or false? Is B true or false?" Once you have completed an individual evaluation of each statement, you will have successfully determined the correct answer choice for the question, "Who is correct?

An important point to remember is that an ASE Technician A, Technician B question will never have Technician A and B directly disagreeing with each other. That is why you must evaluate each statement independently.

An example of a Technician A/Technician B style question looks like this:

1. A noise is heard from the clutch area as the clutch is disengaged. Technician A says that the noise could be a faulty release bearing. Technician B says the noise could be faulty transmission countershaft bearings. Who is correct?

    A. A only
    B. B only
    C. Both A and B
    D. Neither A nor B

TASKS A.1 &
A.4

Answer A is correct. Only Technician A is correct. A faulty release bearing would be most noticeable as the clutch is disengaged.

Answer B is incorrect. The countershaft stops rotating when the clutch is disengaged, so a faulty countershaft bearing would get quieter as the clutch is disengaged.

Answer C is incorrect. Only Technician A is correct.

Answer D is incorrect. Technician A is correct.

EXCEPT QUESTIONS

Another question type used on the ASE exams contains answer choices that are all correct except for one. To help easily identify this type of question, whenever it is presented in an exam, the word "EXCEPT" will always be displayed in capital letters.

Furthermore, a cautionary statement will alert you to the fact that the next question is different from the ones otherwise found in the exam. With the EXCEPT type of question, only one incorrect choice will actually be listed among the options, and that incorrect choice will be the key to the question. That is, the incorrect statement is counted as the correct answer for that question.

Be careful to read these question types slowly and thoroughly; otherwise, you may overlook what the question is actually about and answer the question by selecting the first correct statement.

An example of this type of question would appear as follows:

1. A manual transaxle is being checked for a transaxle oil leak. All of the following could be used to help locate the source of the leak EXCEPT:

    A. A blacklight
    B. White powder
    C. A vacuum gauge
    D. Oil dye

Answer A is incorrect. A blacklight can be used to help locate the source of a leak.

Answer B is incorrect. White powder can be used to help locate the source of a leak.

Answer C is correct. A vacuum gauge is not used to locate transaxle oil leaks.

Answer D is incorrect. Oil dye can be used to help locate a transaxle oil leak.

LEAST LIKELY QUESTIONS

TASK C.1

LEAST LIKELY questions are similar to EXCEPT questions. Look for the answer choice that would be the LEAST LIKELY cause (most incorrect) of the described situation. To help easily identify these type of questions, whenever they are presented in an exam, the words "LEAST LIKELY" will always be displayed in capital letters. In addition, you will be alerted before a LEAST LIKELY question is posed. Read the entire question carefully before choosing your answer.

An example of this type of question is shown below:

1. A rear wheel drive vehicle (RWD) has a grinding noise coming from the differential that only sounds when making turns. Which of the following would be the LEAST LIKELY cause?

    A.   Faulty side gear
    B.   Faulty drive pinion bearing
    C.   Faulty differential pinion gear
    D.   Faulty differential pinion shaft

Answer A is incorrect. A faulty side gear would make noise only on a turn.

Answer B is correct. A faulty drive pinion bearing would make noise going straight as well as on a turn.

Answer C is incorrect. A faulty differential pinion gear would only make noise on a turn.

Answer D is incorrect. A faulty differential pinion shaft would only make noise on a turn.

SUMMARY

TASKS C.13 & C.16

The question styles outlined in this section are the only ones you will encounter on any ASE certification exam. ASE does not use any other types of question styles, such as fill-in-the-blank, true/false, word-matching, or essay. ASE also will not require you to draw diagrams or sketches to support any of your answer selections, although any of the described question styles may include illustrations, charts, or schematics to clarify a question. If a formula or chart is required to answer a question, it will be provided for you.

Task List Overview

INTRODUCTION

This section of the book outlines the content areas or task list for this specific certification exam, along with a written overview of the content covered in the exam.

The task list describes the actual knowledge and skills necessary for a technician to successfully perform the work associated with each skill area. This task list is the fundamental guideline you should use to understand what areas you can expect to be tested on, as well as how each individual area is weighted to include the approximate number of questions you can expect to be given for that area during the ASE certification exam. It is important to note that the number of exam questions for a particular area is to be used as a guideline only. ASE advises that the questions on the exam may not equal the number listed on the task list. The task lists are specifically designed to tell you what ASE expects you to know how to do and to help prepare you to be tested.

Similar to the role this task list will play in regard to the actual ASE exam, Delmar, Cengage Learning has developed six preparation exams, located in Section 5 of this book, using this task list as a guide. It is important to note that although both ASE and Delmar, Cengage Learning use the same task list as a guideline for creating these test questions, none of the test questions you will see in this book will be found in the actual, live ASE exams. This is true for any test preparatory material you use. Real exam questions are only visible during the actual ASE exams.

Task List at a Glance

The Manual Drive Trains and Axles (A3) task list focuses on six core areas, and you can expect to be asked a total of approximately 40 questions on your certification exam, broken out as outlined here:

    A.   Clutch Diagnosis and Repair (6 Questions)

    B.   Transmission Diagnosis and Repair (7 Questions)

    C.   Transaxle Diagnosis and Repair (7 Questions)

    D.   Drive Shaft/Half-Shaft and Universal Joint/Constant Velocity (CV) Joint Diagnosis and Repair (Front- and Rear-Wheel Drive) (5 Questions)

    E.   Rear-Wheel Drive Axle Diagnosis and Repair (7 Questions)

    F.   Four-Wheel Drive/All-Wheel Drive Component Diagnosis and Repair (8 Questions)

Based upon this information, the graph shown here is a general guideline demonstrating which areas will have the most focus on the actual certification exam. This data may help you prioritize your time when preparing for the exam.

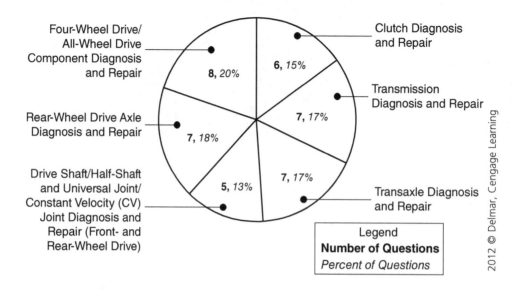

Four-Wheel Drive/
All-Wheel Drive
Component Diagnosis
and Repair — **8**, *20%*

Rear-Wheel Drive Axle
Diagnosis and Repair — **7**, *18%*

Drive Shaft/Half-Shaft
and Universal Joint/
Constant Velocity (CV)
Joint Diagnosis and
Repair (Front- and
Rear-Wheel Drive) — **5**, *13%*

Clutch Diagnosis
and Repair — **6**, *15%*

Transmission
Diagnosis and Repair — **7**, *17%*

Transaxle Diagnosis
and Repair — **7**, *17%*

Legend
Number of Questions
Percent of Questions

2012 © Delmar, Cengage Learning

Note: The actual number of questions you will be given on the ASE certification exam may vary slightly from the information provided in the task list, as exams may contain questions that are included only for statistical research purposes. Don't forget that your answers to these research questions will not affect your score.

MANUAL DRIVE TRAINS AND AXLES (TEST A3) TASK LIST

A. Clutch Diagnosis and Repair (6 Questions)

1. Diagnose clutch noise, binding, slippage, pulsation, chatter, pedal feel/effort, and release problems; determine needed repairs.

Clutch components, the flywheel, pressure plate, and clutch disc, as well as the pilot bearing and release bearing can all contribute to noise, poor shifting, and odd pedal feel/effort.

Either of the bearings can begin to make noise due to lack of lubrication, and can be diagnosed by determining when, during clutch operation, they are quiet and when they are making noise.

Lack of lubrication or rust on the clutch disc splines can cause the clutch disc to bind on the transmission input shaft causing unusual slippage or release problems.

Foreign material such as grease and oil on the clutch disc, or a pressure plate that is beginning to develop weak tension springs can cause slippage of the disc when under power. This foreign material can also cause chattering upon engagement.

As the clutch disc wears and begins to slip, heat checks and small cracks can begin to develop on the flywheel and pressure plate, which will further contribute to slipping and chatter.

Any runout on the flywheel or the transmission mounting surface can cause binding, release problems, pulsation, and chatter. All parts of the clutch mechanism must work together to give a smooth pedal feel/effort.

2. Inspect, adjust, and replace clutch pedal linkage, cables and automatic adjuster mechanisms, brackets, bushings, pivots, springs, and electrical switches.

When there is no clutch pedal free play, the clutch is not fully engaged. The release bearing is touching the fingers of the pressure plate. This will relieve some of the clamping force on the clutch disc, causing the clutch to slip. Hard shifting, improper clutch release, or transaxle gear damage would all be signs of too much clutch pedal free play.

Many late-model vehicles have self-adjusting cables. The cable is adjusted when the clutch pedal is released as the clutch disc wears from normal use. These systems use a constant-running release bearing; it is always in contact with the pressure plate. The clutch pedal will not have any clutch pedal free play.

Worn bushings, pivots, and brackets are the cause of poor clutch engagement and disengagement and should be inspected during clutch inspection.

Springs should also be inspected for wear and replaced as needed.

Electrical switches such as the back-up light switch should be checked for correct operation when diagnosing a no back-up light complaint.

3. Inspect, adjust, replace, and bleed hydraulic clutch slave/release cylinder, master cylinder, lines, and hoses; clean and flush hydraulic system; refill with proper fluid.

The hydraulic system for a hydraulic clutch is totally separate from the brake system. The clutch master cylinder is mounted to the fire wall in the engine compartment. A line runs from the clutch master cylinder to the slave cylinder on the bell housing or inside the bell housing on some models. A rod connected to the clutch pedal goes through the fire wall into the clutch master cylinder. When the rod is pushed into the clutch master cylinder, it forces fluid through the line, which actuates the slave cylinder to release the clutch. The clutch master cylinder push rod should be inspected for moisture contamination, rusting, and pitting.

Air in the system will prevent the clutch from disengaging properly when the clutch pedal is fully depressed.

4. Inspect, adjust, and replace release (throw-out) bearing, bearing retainer, lever, and pivot.

On a clutch with an adjustable linkage, the release bearing should not be in contact with the pressure plate fingers. If the release bearing is not touching the fingers, it will not make any noise even if the bearing is bad. Clutch pedal free play is the distance between the release bearing and the pressure plate fingers; it is the gap or movement in the clutch pedal before the release bearing contacts the pressure plate and releases the clutch.

Hydraulic-controlled clutch systems use a release bearing that is always in contact with the fingers on the pressure plate. There is no manual adjustment on a hydraulic clutch system; it adjusts automatically as the clutch disc wears. When a clutch is disengaged, the release bearing moves toward the pressure plate. The release bearing continues to move toward the pressure plate fingers and compresses the springs in the pressure plate to release the clutch. A worn release bearing retainer, lever, or pivot will affect clutch engagement and disengagement.

5. Inspect and replace clutch disc and pressure plate assembly; inspect input shaft pilot and splines.

Loose or worn engine main bearings may cause an oil leak at the rear main bearing seal, which contaminates the clutch facings with oil, resulting in clutch slippage. An improper pressure plate-to-flywheel position causes engine vibrations.

If the clutch facing is worn too thin, the clamping force of the pressure plate will not be as much as it was when the clutch disc was at full thickness. There will not be enough spring pressure left in the pressure plate to maintain a hard clamping force on the clutch disc because the springs in the pressure plate are fully extended and not applying enough pressure.

Proper installation of the clutch disc is critical; the damper spring offset must face the transmission. The clutch disc is normally marked to indicate which side should face the flywheel. If the damper spring offset is toward the engine, the springs may contact the flywheel or the flywheel bolts, damaging these components.

Inspect the shaft tip that rides in the pilot bearing for smoothness. Check the splines of the shaft for any wear that could prevent the clutch disc from sliding evenly and smoothly. If the splines have excessive wear or damage, this will cause the clutch to engage roughly. A clutch disc that is bent, or that has weak torsional springs, will also cause the clutch to engage roughly.

6. Inspect pilot bearing/bushing inner and outer bores; inspect and replace pilot bearing/bushing.

The pilot bearing supports the forward end of the transmission input shaft and keeps the clutch disc, which is riding on that shaft, in alignment with the flywheel and pressure plate.

When the clutch is engaged, the transmission input shaft rotates at the same speed as the engine flywheel and pilot bearing at all times. When the clutch is disengaged, the flywheel and pilot bearing rotate on the end of the transmission input shaft and turn faster than the shaft.

If a bushing-type pilot bearing (iolite) is lubricated with bearing grease, friction actually will increase between the bushing and the transmission input shaft due to the grease blocking the release of the oil-impregnated bushing. Lubricate a bushing-type pilot bearing with motor oil. Lubricate a roller-type pilot bearing with wheel bearing grease.

7. Inspect and measure flywheel and ring gear; inspect dual-mass flywheel where required; repair or replace as necessary.

If too much material is removed from the flywheel, the torsion springs on the clutch plate are moved closer to the mounting bolts on the flywheel, and these springs may contact the heads of the flywheel bolts. Removing excessive material from the flywheel moves the pressure plate forward, away from the release bearing. This action increases free play, so the slave cylinder rod may not move far enough to release the clutch.

If the flywheel is not resurfaced, damage or premature wear can be caused to the rest of the new assembly that was installed. A flywheel should be resurfaced every time a clutch assembly is replaced. Resurfacing the flywheel will ensure that it has the flatness and microfinish it needs to ensure that the new clutch disc breaks in properly. If it is not resurfaced, the clutch disc will probably glaze and chatter. Clean the flywheel with hot water and soap to remove all residue on the surface after it has been resurfaced.

Designs of dual-mass flywheels vary, so it is most important to know what they do. A dual-mass flywheel is designed to isolate torsional crankshaft spikes created by diesel engines with high compression ratios. With the separation of the mass of the flywheel between the engine and the transmission, torsional spikes can be isolated, eliminating potential damage to the transmission gear teeth. You will need to inspect the springs or other dampening method used to connect the two separate components of the flywheel.

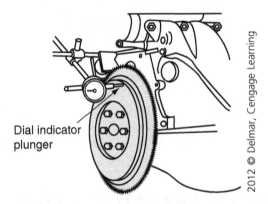

Dial indicator plunger

2012 © Delmar, Cengage Learning

8. Inspect engine block, clutch (bell) housing, transmission case mating surfaces, and alignment dowels; inspect engine core plugs, rear main engine seal, and other sources of fluid contamination; determine needed repairs.

If the bell housing is not aligned properly with the engine block because of something being pinched between them, or because of a burr or imperfection on one of the mating surfaces, worn alignment dowels, or loose bell housing bolts, the clutch will not make even contact with the flywheel. This will cause the clutch to chatter and grab because of the uneven contact when the clutch is being engaged.

If a clutch disc or an input shaft is bent due to careless removal or installation of the transmission, this will cause misalignment problems. If the transmission is misaligned when it is installed, this can cause the pilot bushing or bearing to wear out prematurely.

The rear main seal should be inspected when the transmission is removed for any signs of leakage and replaced at this time; this is also the time to inspect core plugs for any sign of leakage. Leakage from the rear main, core plug, or any other component can contaminate the clutch disc and cause premature failure.

9. Measure flywheel surface runout and crankshaft end-play; determine needed repairs.

To measure crankshaft end-play, mount the magnetic dial indicator on the back of the engine block. Position the dial indicator to the flywheel. Push the flywheel toward the front of the engine until it stops. Adjust the dial indicator to zero and then pull the flywheel toward the back of the engine. The reading on the dial indicator will be the crankshaft end-play.

Excessive main bearing wear will cause low oil pressure or a rear main oil leak, possibly causing the clutch disc to become contaminated with engine oil. A thrust bearing is placed between the crankshaft main bearing cap and the side of a crankshaft journal. These thrust bearings are put into place to control the forward and rearward movements of the crankshaft during acceleration and deceleration. The proper thickness thrust bearing is selected when the engine is assembled to set the crankshaft end-play.

10. Inspect, replace, and align power train mounts.

Engine and transmission mounts should be inspected for broken, sagging, oil-soaked, or deteriorated conditions. Any of these mount conditions may cause a grabbing, binding clutch. On a rear-wheel drive car, damaged engine or transmission mounts may cause improper drive shaft angles, which result in a vibration that changes in intensity when the vehicle accelerates and decelerates.

On a vehicle with a manual clutch, part of the clutch linkage is connected from the frame of the vehicle to the engine block. If the vehicle has a broken engine mount, the engine can lift off the frame on acceleration. This can cause the clutch linkage to move and apply release pressure to the release bearing fork. This may cause the release bearing to apply release pressure on the pressure plate, causing the clutch disc to slip between the flywheel and the pressure plate.

B. Transmission Diagnosis and Repair (7 Questions)

1. Diagnose transmission noise, difficult shifting, gear clash, jumping out of gear, fluid condition and type, and fluid leakage problems; determine needed repairs.

A misaligned or loose transmission case or clutch housing may cause the transmission to slip out of gear. Check the mounting of the transmission to the engine block for looseness or dirt between the two cases. Broken or loose engine mounts will also cause misalignment. This problem may also be caused by insufficient spring tension of the shift rail detent spring or bent or worn shift forks, levers, or shafts. Other internal problems may cause this condition, such as a worn input shaft pilot bearing, bent output shaft, or a worn or broken synchronizer. Transmission bearing noise is the cause of many transmission related complaints. The transmission should be operated through each of its gears and close attention must be paid to any change in noise; this can help locate the bad bearings before disassembling.

If the transmission shifts hard or the gears clash while shifting, a common cause of the problem is the clutch. Check the clutch pedal free-travel adjustment. Make sure the clutch releases completely. Also check for worn clutch parts and a binding input shaft pilot bearing. Shift linkage problems can also cause this problem. If the shift lever is worn, binding, or out of adjustment, proper engagement of gears is impossible. An unlubricated linkage will also cause shifting difficulties.

A transmission can use automatic transmission fluid, gear oil, motor oil, and manual transmission fluid depending on the make and model. The correct type of fluid must be used or hard shifting and leaks can be caused. If a leak is found and the correct fluid is used, the transmission should be cleaned and dye should be used to locate the location of the leak.

2. Inspect, adjust, and replace transmission external shifter assembly, shift linkages, brackets, bushings/grommets, pivots, and levers.

Most external shift linkages and cables require adjusting; this usually is performed by placing the transmission in neutral and the shifter in the neutral position. Raise the vehicle and place the shifter in the neutral position to begin the shift linkage adjustment. With a lever-type shift linkage, install a rod in the adjustment hole in the shifter assembly. Adjust the shift linkages by loosening the rod retaining locknuts and moving the levers until the rod fully enters the alignment holes. Tighten the locknuts and check the shift operation in all gears.

Transmissions with internal linkage do not have provisions for adjustment. However, external linkages, both floor- and column-mount, can be adjusted. Linkages are adjusted at the factory, but worn parts may make adjustment necessary. Also, after a transmission has been disassembled, the shift lever may need adjustment.

3. Inspect and replace transmission gaskets, sealants, seals, and fasteners; inspect sealing surfaces.

Excessive output shaft or excessive input shaft end-play results in lateral shaft movement that may adversely affect the extension housing seal. A worn output shaft bushing will cause premature repeat extension housing seal failure. If the drive shaft yoke has a score or imperfection on the shaft, it could damage the seal and cause the transmission to leak fluid at the rear of the transmission. Replacing the seal will not correct this condition unless the yoke is replaced.

A cork gasket should be installed as it is when it comes out of the box. It was made to be installed dry and does not require any type of added sealant to help the gasket seal any better. A spray adhesive sometimes may be used to hold the gasket in place to help installation. A rubber gasket should not use any additional gasket sealant when installed; it will become too slippery and may not position correctly when being installed. Some transmissions use room temperature vulcanizing (RTV), which means it cures in the presence of air. Some transmissions use an anaerobic sealer that cures in the absence of air. Extra care should be used to be sure the correct length fastener is used. Installing a fastener that is too long can cause noise or even lock up the transmission.

4. Remove and replace transmission; inspect transmission mounts.

Care must be exercised when removing and installing the transmission to avoid damaging components. If the transmission is supported from the input shaft, the weight of the transmission itself could cause the input shaft and the clutch disc to be bent or damaged. The engine support fixture must be installed before the transmission-to-engine bolts are even loosened. When the transmission is being removed, the clutch disc may move, causing it to be misaligned. A clutch disc alignment tool must be used before the transmission is installed to align the clutch disc with the flywheel.

A transmission mount absorbs a lot of torque and vibration in the rubber mount when shifting and accelerating. If the rubber mount becomes saturated with oil, the oil will deteriorate the rubber and weaken it. This will eventually cause the mount to fail. Oil will not cause the mount to crack, but the oil will make the rubber in the mount feel soft and spongy.

The drive shaft should be installed in the original position on the rear axle pinion flange. If it is not installed in the original position, you may experience a driveline vibration.

5. Disassemble and clean transmission components; reassemble transmission.

The first step in disassembling a transmission is to obtain the correct service manual for the transmission being disassembled. Each step should be followed closely to prevent damage to the transmission.

Clean the transmission case with a steam cleaner, degreaser, or cleaning solvent. As you begin to disassemble the unit, pay close attention to the condition of its parts. Using a dial indicator, measure and record the end-play of the input and mainshafts. This information will be needed during assembly of the unit to select the appropriate selective shims and washers. The transmission parts should be cleaned with solvent, dried, and lubricated before assembly.

6. Inspect, repair, and/or replace transmission shift cover and internal shift forks, bushings, bearings, levers, shafts, sleeves, detent mechanisms, interlocks, and springs.

Shift rails should be inspected to be sure that they are not bent. A bent rail may cause hard shifting. The shift rail is linked to the shifter handle and if the rail is bent, it may interfere with other parts in the transmission, causing the hard shift feeling when the shift lever is moved from certain gears.

A linkage that is bent and a linkage that is out of adjustment are both common causes of a transmission that does not shift properly. Failure to go into gear is more commonly caused by a broken shift fork than by a damaged gear.

The shift linkages internal to the transmission are located at the top or side of the housing. Mounted inside the transmission is the control end of the shifter and the shift controls. The shift controls consist of the shift rail and the shift fork. As the shift fork moves toward the preferred gear, it moves the synchronizer sleeve to lock the speed gear to the shaft.

7. Inspect and replace input (clutch) shaft, bearings, and retainers.

The input shaft should be inspected for wearing the splines and the pilot bearing area. The bearing should be checked for smooth operation and the bearing retainer should be inspected for wear especially on the end where the release bearing rides.

8. Inspect and replace mainshaft, gears, thrust washers, bearings, and retainers/snap rings; measure clearance and end-play.

The mainshaft supports many of the drive gears as well as the synchronizer assemblies, and has a bearing surface at the forward end as well as a support bearing at the rear.

The mainshaft is generally not drilled with oil journals. Inspect the bearing surfaces of the mainshaft; it should be smooth and show no signs of overheating. Also inspect the gear journal areas on the shaft for roughness, scoring, and other defects. Check the shaft splines for wear, burrs, and other conditions that would interfere with the slip yoke's ability to slide smoothly on the splines.

Each gear on the mainshaft should be inspected for damaged drive teeth, damaged dog teeth, and scored or overheated journal surfaces.

Thrust washers and retainers/snap rings must be measured for proper clearance, and an end-play check performed with a dial indicator and compared to manufacturer's specs.

9. Inspect and replace synchronizer hub, sleeve, keys (inserts), springs, and blocking (synchronizing) rings/mechanisms; measure blocking ring clearance.

Due to the speed with which shifts are made in a transmission, great care must be taken in inspecting each part of the synchronizer assembly to ensure smooth shifting.

The blocking ring dog teeth tips should be pointed with smooth surfaces. Clearance between the blocking ring and the matching gear dog teeth is important for proper shifting. For example, if the clearance between the blocking ring and the fourth-speed gear dog teeth is less than specified, the blocking ring is worn. This limits its clutching ability, resulting in hard shifting or grinding. The threads on the blocking ring in the cone area must be sharp to get a good bite on the gear to stop it from spinning, and to make a synchronized non-clashing shift.

The synchronizer sleeve must slide freely on the synchronizer hub. Due to wear mating issues between the hub and the sleeve over time, it is important to mark them before disassembly to ensure they are reassembled on the same splines.

10. Inspect and replace countershaft, counter (cluster) gear, bearings, thrust washers, and retainers/snap rings.

The counter (cluster) gear is located in the lower part of the transmission and is in mesh with the speed gears located on the mainshaft. Since the counter (cluster) gear is turning with the clutch pedal released in neutral and in gear, damaged counter (cluster) gear bearings may cause a growling noise with the engine idling with the transmission in neutral and the clutch pedal released. They will also cause a growling noise while driving in any gear, but tend to be noisiest in the lower gears.

All counter (cluster) gears should show wear patterns in the center of their teeth. These wear patterns should appear as a polished finish, with little wear on the gear face. Check the gear's teeth carefully for chips, pitting, cracks, or breakage. Also, inspect the bearing surfaces to make sure they are smooth. Any damage to the assembly requires replacement.

11. Inspect and replace reverse idler gear, shaft, bearings, thrust washers, and retainers/snap rings.

Since the reverse idler gear is only in mesh with reverse gear, this gear rotates in reverse only. Some transmissions actually turn the reverse idler gear all the time and mesh the reverse gear with the idler, or some five-speeds use a synchronizer to connect a constantly engaged idler and its gear to the mainshaft.

Inspect the reverse idler gear for pitted, cracked, nicked, or broken teeth. Check its center bore for a smooth surface. Carefully inspect the needle bearings on which the idler gear rides for wear, burrs, and other defects. Also, inspect the reverse idler gear's shaft surface for scoring, wear, and other imperfections. Replace any part that is damaged or excessively worn.

12. Measure and adjust bearing preload or end-play (shim/spacer selection procedure).

Excessive input shaft or mainshaft end-play can cause the transmission to jump out of gear. For example, in fourth gear, the 3-4 synchronizer is moved ahead so the synchronizer sleeve is meshed with the dog teeth on the fourth-speed gear on the input shaft. While disassembling a transmission, you should be taking end-play readings. These readings will be recorded, and when the transmission is reassembled, select-fit thrust washers and shims will be used to set all parts to specifications. Bearings and snap rings can be reused if nothing is found during cleaning and inspection.

13. Inspect, repair, and replace extension housing and transmission case mating surfaces, bores, dowels, bushings, and vents.

Check the extension housing for cracks and repair or replace it as needed. Check the mating surfaces of the housing for burrs or gouges and file the surface flat. Inspect the speedometer cable or sensor for any leakage from the seal; replace the seal if fluid is leaking. Install a new gasket to the extension housing during installation. Check all threaded holes and repair any damaged bores with a thread repair kit. Check the bushing in the rear of the extension housing for excessive wear or damage. Always replace the rear extension housing seal.

14. Inspect and replace transmission components related to speedometer operation.

The speedometer driving gear is the gear located on the output shaft of the transmission and can be accessed by removing the rear extension housing. The speedometer driven gear is located on the vehicle speed sensor and is located in the rear extension housing.

15. Inspect, test, and replace transmission sensors, actuators, and switches.

Manual transmissions may have various sensors, switches, and solenoids. These components may include a vehicle speed sensor (VSS), a back-up light switch, and a computer-aided gear select solenoid. The VSS sends a voltage signal to the power train control module (PCM) in relation to vehicle speed. The PCM uses this signal to operate output such as the cruise control. Some vehicles have multiplex sensors and obtain the VSS signal from antilock braking system (ABS) wheel speed sensors. The ABS/VSS sensor is usually a permanent-magnet signal generator. However, some units utilize a Hall-effect device to generate the voltage signal. While some of these types of sensors may be checked using an ohmmeter for initial resistance, the only accurate method of testing the sensors is by using a digital multi-meter (DMM) on the AC voltage setting or by using an oscilloscope.

The back-up light switch is operated by linkage inside the transmissions to operate the back-up lights when the transmission is placed in reverse. The switch is normally open and has power going to it when the vehicle ignition is on. The back-up lamp lights up when the vehicle is shifted into reverse. The switch will close when the vehicle is shifted in reverse and provide a path to ground for the back-up lamps to operate. The computer-aided gear select solenoid ensures good fuel economy and compliance with federal fuel economy standards by inhibiting second and third gears when shifting out of first gear under certain conditions.

16. Inspect lubrication systems.

Transmission fluid level must be maintained at the level of the check plug in the transmission case or at a level marked on a transmission dipstick. Many late model manual transmissions and transaxles use automatic transmission fluid (ATF) as a lubricant for reduced friction and improved vehicle fuel economy. Some manual transmissions use hypoid gear oil as a lubricant; some use motor oil.

The hypoid ring-and-pinion gear sets in rear-wheel drive axles require hypoid gear oil, usually GL4 or GL5. Limited slip differentials require additional fluid additives. The viscosity of hypoid gear oil is higher (thicker) than that of motor oil or ATF. It may be single-viscosity, such as SAE 90, or multiple-viscosity, such as 85W-90. Many final-drive gear sets in front-wheel drive (4WD) transaxles are not hypoid gears and use ATF or motor oil as a lubricant. Some 4WD final drives are hypoid gear sets however, and require GL4 or GL5 gear oil. Always follow the manufacturer's specifications for fluid type, viscosity, and replacement intervals.

17. Check fluid level; refill with proper fluid.

It is important for technicians to include an inspection of drive train fluid level and to be sure that manufacturer's recommended types of fluids are used. As a technician, you should check manufacturer's technical service bulletins (TSBs) for changes in fluids or additives that have been recommended since the production date. When fluid level is incorrect, it should be adjusted with the correct fluid type to proper levels.

C. Transaxle Diagnosis and Repair (7 Questions)

1. Diagnose transaxle noise, difficult shifting, gear clash, jumping out of gear, fluid condition and type, and fluid leakage problems; determine needed repairs.

A misaligned or loose transaxle case or clutch housing may cause the transaxle to slip out of gear. Check the mounting of the transaxle to the engine block for looseness or dirt between the two cases. Broken or loose engine mounts will also cause misalignment. This problem may also be caused by insufficient spring tension of the shift rail detent spring or bent or worn shift forks, levers, or shafts. Other internal problems may cause this condition, such as a worn input shaft pilot bearing, bent output shaft, or a worn or broken synchronizer. Transaxle bearing noise is the cause of many transaxle related complaints. The transaxle should be operated through each of its gears and close attention should be paid to any change in noise; this can help locate the bad bearings before disassembling.

If the transaxle shifts hard or the gears clash while shifting, a common cause of the problem is the clutch. Check the clutch pedal free-travel adjustment. Make sure the clutch releases completely. Also check for worn clutch parts and a binding input shaft pilot bearing. Shift linkage problems can also cause this problem. If the shift lever is worn, binding, or out of adjustment, proper engagement of gears is impossible. An unlubricated linkage will also cause shifting difficulties.

A transaxle can use automatic transaxle fluid, gear oil, motor oil, and manual transaxle fluid depending on the make and model. The correct type of fluid must be used or hard shifting and leaks can be caused. If a leak is found and the correct fluid is used, the transaxle should be cleaned and dye should be used to locate the location of the leak.

2. Inspect, adjust, lubricate, and replace transaxle external shift assemblies, linkages, brackets, bushings/grommets, cables, pivots, and levers.

A misadjusted shift linkage may cause many problems. If the linkage is misadjusted, the transaxle may not be able to be shifted all the way into gear. This will cause further damage to the transaxle. An improper shift linkage adjustment may also cause hard shifting or sticking in gear.

Most transmissions and transaxles are adjusted with the unit in neutral. A 1/4-inch (6.35 mm) bar or drill bit is installed in the lever to hold the transaxle in neutral while the cables or linkage are adjusted. Transmissions and transaxles with internal linkage have no adjustment. Shift cables should not be modified in any way, only replaced or adjusted.

3. Inspect and replace transaxle gaskets, sealants, seals, and fasteners; inspect sealing surfaces.

Transaxle cases have machined mating surfaces that have a very smooth flat finish on them. Not all require a gasket, but they do require some sort of sealant that should be equivalent to the manufacturer's specifications. If a transaxle is assembled without following manufacturer's sealing instructions, leakage from the case will result.

The transaxle mating surfaces should be inspected for warpage with a straightedge before assembly to ensure a proper fit.

A plugged transaxle vent may cause excessive transaxle pressure and repeated drive axle seal failure.

4. Remove and replace transaxle; inspect, replace, and align transaxle mounts and subframe/cradle assembly.

A misaligned engine and transaxle cradle may cause drive axle vibrations. Since the lower control arms are connected to this cradle, misalignment of the cradle may cause improper front suspension angles.

When removing a transaxle from a vehicle, it is necessary to install an engine support bracket. This will hold the weight of the engine while the transaxle is being removed. It is installed for the safety of the technician, as well as to avoid damaging the vehicle. You do not have to drain engine oil when removing a transaxle. You should disconnect the negative battery cable. Not all vehicles require you to remove the engine along with the transaxle.

5. Disassemble and clean transaxle components; reassemble transaxle.

Clean the transaxle case with a steam cleaner, degreaser, or cleaning solvent. As you begin to disassemble the unit, pay close attention to the condition of its parts. Using a dial indicator, measure and record the end play of the appropriate shafts. This information will be needed during assembly of the unit to select the appropriate selective shims and washers.

Before disassembling a transaxle, observe the effort it takes to rotate the input shaft through all forward gears and reverse. Extreme effort in any or all gears may indicate an end-play problem or a bent shaft.

While assembling a manual transaxle, it is important to apply gear lube to all of the transaxle parts. Before checking the specifications of the shafts in the transaxle, rotate the shafts to work the gear lube into the bearings. If the gear lube is not worked into the bearings, a false measurement may be made.

6. Inspect, repair, and/or replace transaxle shift cover and internal shift forks, levers, bushings, shafts, sleeves, detent mechanisms, interlocks, and springs.

Shift rails should be inspected to be sure that they are not bent. Hard shifting may be a cause of a bent shift rail. The shift rail is linked to the shifter handle and if the rail is bent, it may interfere with other parts in the transmission, causing the hard shift feeling when the shift lever is moved from certain gears.

A linkage that is bent and a linkage that is out of adjustment are both common causes of a transmission that does not shift properly. Failure to go into gear is more commonly caused by a broken shift fork than by a damaged gear.

The shift linkages internal to the transmission are located at the top or side of the housing. Mounted inside the transmission is the control end of the shifter and the shift controls. The shift controls consist of the shift rail and the shift fork. As the shift fork moves toward the preferred gear, it moves the synchronizer sleeve to lock the speed gear to the shaft.

7. Inspect and replace input shaft and output shaft, gears, thrust washers, bearings, and retainers/snap rings.

The input shaft should be inspected for wear at the splines and the pilot bearing area if a pilot bearing is used. The bearings should be checked for smooth operation and the gears and retainers inspected. Snap rings should not be reused.

8. Inspect and replace synchronizer hub, sleeve, keys (inserts), springs, and blocking (synchronizing) rings; measure blocking ring clearance.

Before disassembly, always mark the synchronizer sleeve and hub so that these components can be reassembled in their original locations. Synchronizer hubs are not reversible on the shaft, and synchronizer sleeves are not reversible on their hubs.

A worn blocker ring will cause the transmission/transaxle to have gear clash. If the blocker ring is worn in the cone area (meaning that all of the sharp ridges are dull or gone), the blocker ring will not work properly. A blocker ring should stop a gear from spinning through the sharp ridges in the cone area before the synchronizer sleeve engages the gear. As the blocker ring wears, the blocker ring moves closer to the gear; this wear is checked by measuring the clearance with a feeler gauge.

9. Inspect and replace reverse idler gear, shaft, bearings, thrust washers, and retainers/snap rings.

Inspect the reverse idler gear teeth for chips, pits, and cracks. Although worn reverse idler teeth may cause a growling noise while driving in reverse, this problem would not cause a failure to shift into reverse. The needle bearings on a reverse idler gear should be smooth and shiny. Carefully inspect the needle bearings for wear, burrs, and other problems. Worn or damaged needle bearings should be replaced, or damage to other components may occur.

10. Inspect, repair, and/or replace transaxle case mating surfaces, bores, dowels, bushings, bearings, and vents.

Transaxle case replacement is often required if the case is cracked. Some vehicle manufacturers recommend that the case be repaired, depending on how extensive the damage. The transaxle case may be repaired with an epoxy-based sealer for some transaxle cracks. Loctite® is not recommended for repairing any transaxle case.

If a threaded area in an aluminum housing is damaged, service kits can be used to insert new threads in the bore. Some threads should never be repaired; check the service manual to identify which ones can be repaired. When bushing and bearing replacement is necessary, the correct driver should be used for correct installation. A plugged transaxle vent can cause the transaxle to build excess pressure and fluid leakage.

11. Inspect and replace transaxle components related to speedometer operation.

A speedometer gear is normally mounted on the output shaft. The output shaft spins at driveline speed. An output shaft will never have a drive gear machined into the output shaft. A drive gear is made out of a plastic nylon-type gear, and the teeth on the drive and driven gear have a helical cut on them. The helical cut and the plastic-type gear are used for quiet operation.

Whenever a speedometer drive gear is replaced, the driven gear should also be replaced. If a speedometer cable assembly core is damaged, it may be replaced with a new core. The new core must be cut to the same size as the one being replaced and be properly lubricated before installation into the cable housing.

12. Inspect, test, and replace transaxle sensors, actuators, and switches.

Most VSSs are magnetic pick-up coil signal generators. Some from the early 1980s, however, are rotary magnetic switches or optical sensors. The signal from a magnetic pick-up VSS is an analog sine wave that varies in frequency and amplitude with vehicle speed. The signal from a magnetic switch or an optical VSS is a digital square wave that varies in frequency only. The signal integrity and waveform of any kind of VSS are best tested with an oscilloscope.

13. Diagnose differential assembly noise and wear; determine needed repairs.

Damaged ring gear teeth would cause a clicking noise while the vehicle is in motion. This problem would not cause differential chatter. Improper preload on differential components, such as side bearings, may cause differential chatter.

If there is a constant whining noise coming from the differential, the noise cannot be coming from the side gears or the spider gears. These gears are used only when the vehicle is turning, so if they were damaged, the noise would only be heard on turns. The wrong differential lube could cause damage to the differential parts but will not cause a whining noise. If the preload and backlash are not set properly, the gear mesh could be too tight and cause a whining noise.

14. Remove and replace differential final drive assembly.

In most transaxles, the differential is serviced or removed by disassembling the halves of the transaxle case. The differential bearings fit into the case and are not usually adjustable for bearing preload. If the side bearings allow for preload, there are shims that fit between the transaxle housing and the bearing cup or assembly. The differential can usually be removed without complete disassembly of the transaxle gear train.

15. Inspect, measure, adjust, and replace differential pinion (spider) gears, shaft, side gears, thrust washers, side bearings, and case/carrier.

The side gear end-play must be measured individually on each side gear with the thrust washers removed. Side gears with the specified thrust washer have slight end-play, but no preload.

The spider gears ride on the pinion shaft, and the bore in the gears should be smooth and shiny and have no signs of pits or scuffing. The pinion shaft should also be free of pitting and scoring. There is no needle bearing in any of the spider gears.

16. Diagnose limited slip differential noise, slippage, and chatter problems; determine needed repairs.

Slipping or worn fiber discs in a limited slip differential will cause a loss of traction through the limited slip differential. Not all limited slip differentials use fiber discs. Some use special tapered cone-type side gears with preload springs causing outward pressure on the side gears. Some even use a gerotor oil pump that is driven when axles operate at different speeds. Most limited slip differentials require that a special additive be added to the gear oil; this additive is a friction modifier. Premature wear causing slipping and/or chatter can happen without this additive. Limited slip differentials are checked with one wheel off the ground and the use of a torque wrench to measure rolling resistance.

17. Measure and adjust shaft and differential bearing preload and end-play (shim/spacer selection procedure).

While you measure the differential side-play to determine the required side bearing shim thickness, a new bearing cup is installed in the transaxle case without the shim. The proper shim thickness is equal to the differential end-play plus a specified thickness for bearing preload. While you measure the differential end-play, the transaxle case bolts must be tightened to the specified torque. A medium load should be applied to the differential in the upward direction while measuring the differential end-play.

When the differential turning torque is less than specified, the shim thickness must be increased behind the side bearing cup in the bell housing side of the transaxle.

18. Inspect lubrication systems.

Wrong transaxle lubricant may cause burned output shaft bearings. A broken oil feeder behind the front output shaft bearing results in improper output shaft bearing lubrication and burned bearings. If a bearing is not lubricated properly, heat will build and destroy the bearing and possibly other parts of the transaxle.

During transmission or transaxle overhaul, drain and inspect the fluid. Gold-colored particles in the fluid are from the wearing of the brass blocking rings on the synchronizers. Metal shavings in the fluid are produced from the wearing of the gears. An excessive amount of shavings in the fluid indicates severe gear and synchronizer wear.

On a cold start, if the fluid is too thick, the vehicle may be hard to shift. This will also cause the transaxle to get poor lubrication on a cold start.

19. Check fluid level; refill with proper fluid.

It is important for technicians to include an inspection of drive train fluid level and to be sure that manufacturer's recommended types of fluids are used. As a technician, you should check manufacturer's TSBs for changes in fluids or additives that have been recommended since the production date. When fluid level is incorrect, it should be adjusted with the correct fluid type to proper levels.

D. Drive Shaft/Half-Shaft and Universal Joint/Constant Velocity (CV) Joint Diagnosis and Repair (Front- and Rear-Wheel Drive) (5 Questions)

1. Diagnose shaft and universal/CV joint noise and vibration problems; determine needed repairs.

A worn universal joint (U-joint) may cause a squeaking noise that increases in relation to vehicle speed. Often, rust will be visible around the grease seals of the caps as well.

If the centering ball and socket are worn in a double cardan U-joint, a heavy vibration may occur during acceleration.

If the torsional damper is bad, a shudder will be felt in the vehicle.

Half-shaft CV joint problems will show up differently depending on whether the inner or outer joint is at fault. For example, a CV inner joint is not affected when the vehicle is turning, it is only affected when the vehicle suspension is jounced or rebounded while driving. Sometimes a clunking noise may be heard upon acceleration or deceleration, while an outer CV joint will make a clicking noise when the vehicle is turning. In both cases, this means the joint ball bearings are bad and the grease is contaminated. In this case, the CV joint should be replaced. As a general rule, axle shafts are serviceable and therefore the whole axle shaft does not need to be replaced.

2. Inspect, service, and replace shafts, yokes, boots, and universal/CV joints; verify proper phasing.

A universal joint has pre-lube on the inside, but this pre-lube is not sufficient to lubricate the joint when it has been installed in a vehicle. You should grease a new universal joint when it is installed. When grease is seen purging at each cap, it ensures that the grease has reached the joint end. This can prevent damage to the trunnion due to lack of lubrication, which could shorten the life of the joint. A number one cause of universal joint failure is brinelling. If a two-piece drive shaft is used, the drive shaft must be timed or phased correctly to prevent vibration. Some slip yokes use a boot to prevent dirt and water from entering and to keep lube in. If a boot is found worn, it must be replaced.

3. Inspect, service, and replace shaft center support bearings.

A worn drive shaft center bearing causes a growling noise that is not influenced by acceleration and deceleration.

A center support bearing is usually maintenance-free, and is typically seated in a rubber support that must be inspected for looseness and deterioration. A center support bearing is found mainly on trucks and vans and is used to break up the length of long drive shafts into two sections to decrease pinion angles.

4. Check and correct drive/propeller shaft balance.

Before checking drive shaft balance, always inspect the shaft for damage. A missing balance weight, accumulation of dirt, or excessive undercoating will affect the balance.

To check drive shaft balance, chalk mark the drive shaft at four locations 90 degrees apart slightly in front of the drive shaft balance weight. Hold a strobe light against the rear axle housing just behind the pinion yoke. Run the vehicle in gear until the drive shaft vibration is the most severe. Point the strobe light at the chalk marks on the drive shaft and note the position of one reference mark. The number that appeared on the strobe light should have two screw-type clamps installed on the shaft near the rear with the heads of the clamp opposite the number that appeared. The vehicle's suspension should have the weight of the vehicle on it when this procedure is performed so that the suspension is at its normal ride height and there are no abnormal pinion angles.

5. Measure drive shaft runout.

While you measure drive shaft runout, the dial indicator should be positioned near the center of the drive shaft. The drive shaft should be replaced if the runout is excessive. A drive shaft that is bent or damaged in any way should be replaced; repairs to a damaged drive shaft should not be attempted.

To check drive shaft runout, raise the vehicle and install a dial indicator with a magnetic base under the vehicle near the center of the drive shaft. The surface of the shaft should be wiped off or cleaned in case it is rusted or has dirt buildup that may affect the reading on the dial indicator. Rolling a drive shaft on a flat surface is not an accurate way of checking the drive shaft runout.

6. Measure and adjust drive shaft working angles.

Drive shaft working angle is also known as pinion angle. The engine and transmission are installed in the chassis at a preset angle, usually with the driveline pointing down at the tail shaft in rear-wheel drive applications. For our purposes, let's assume that the tail shaft end of the driveline is pointing down 3 degrees. To avoid a humming or droning vibration in the driveline, the pinion end of the driveshaft must be pointing up 3 degrees. This creates a situation where the planes that they operate in are parallel and provides the smoothest U-joint operation. (Note that most manufacturers only give specs for pickup trucks for this measurement. Unless the manufacturer specifies differently, the static setting of each should be the same.) Causes of drive shaft working angle problems are sagging (damaged) rear springs, sagging engine or transmission mounts, or major changes in ride height up or down.

7. Inspect, service, and replace wheel bearings, seals, and hubs.

Wheel bearings should be cleaned, inspected for wear and damage, and packed with the appropriate wheel bearing grease using either a packing tool or the hand-pack method.

If only one bearing is bad, replace only that bearing. When a wheel bearing is replaced, the bearing race should also be replaced. The new bearing may fail if the race is not replaced.

E. Rear-Wheel Drive Axle Diagnosis and Repair (7 Questions)

1. Ring and Pinion Gears (3 Questions)

1. Diagnose noise, vibration, and fluid leakage problems; determine needed repairs.

Noises in differentials can be from the axle bearings, ring and pinion, or any of the differential bearings or differential side gears. Axle bearings usually can be isolated by the change in noise as the vehicle experiences different side loads or by raising the vehicle and running it in gear. Most axle bearing noises will subside dramatically when weight is taken off the wheels. Ring and pinion noises are associated with a whine or growl that changes in pitch as vehicle speed or engine load changes. To diagnose differential carrier and pinion bearing noises, a stethoscope is often employed to pinpoint the location of the noise. Since the side gears are only turning while cornering, they do not cause a noise while the vehicle is driven straight ahead. If the differential fluid is too full, excessive pressure may build up and cause the differential fluid to leak past a seal. If the vent becomes plugged, it will cause excessive pressure in the differential housing and a leak will occur. Axle shaft bearings that are worn will cause the axle shaft to apply load on the axle shaft seal, and the seal will fail. Pinion seals and carrier covers are some other leak points that are common and easy to spot.

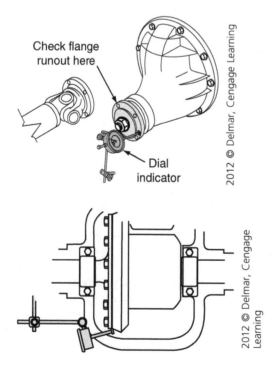

Check flange runout here

Dial indicator

2012 © Delmar, Cengage Learning

2012 © Delmar, Cengage Learning

2. Inspect and replace companion flange, yoke, and pinion seal; measure companion flange runout.

To check pinion flange runout, remove the drive shaft and mount a dial indicator against the face of the flange, as shown in the illustration. Rotate the flange and note the readings on the dial indicator; these are the runout readings. If the flange is removed, you cannot measure pinion flange runout. A loose pinion nut allows pinion shaft end-play, resulting in a clunking noise on acceleration and deceleration. Insufficient pinion nut torque will affect the pinion bearing and may cause pinion seal leakage.

3. Measure ring gear runout; determine needed repairs.

Ring gear runout is best measured before differential disassembly. In most applications, a dial indicator is mounted to the carrier or axle housing and the measuring tip is set on the back of the ring gear (the side without gear teeth). Runout is measured by turning the ring gear and watching for variations in the dial indicator. Use manufacturer specs to determine if the gear has excessive runout. Possible causes for ring gear distortion are uneven torque of ring gear bolts, overheating of gear assembly (usually evident long before you take this measurement), manufacturing defects, assembly flaws or accidents, and debris between the carrier and the gear. Torque and debris issues can usually be corrected without replacing the gear, but all other issues require ring and pinion set replacement.

4. Inspect and replace ring and pinion gear set, collapsible spacers, sleeves (shims), and bearings.

This task covers an enormous area. Much of the related information in this task is in other areas of this section of the Task List. If you are not familiar with the process of rebuilding a differential, it would be wise to read through any manufacturer's procedure, as it exceeds the scope of this book. Key issues that you should know: The ring and pinion set must be replaced as a set, collapsible spacers must be replaced if they are over tightened or when the pinion bearings need to be replaced, and bearings should be replaced as assemblies. When determining the condition of a ring and pinion, you are looking for excessive wear on the gears, pitting or grooves worn in the gear faces, or evidence that the gears have overheated, which will cause them to turn blue or black.

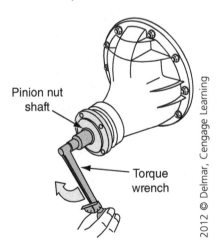

Pinion nut
shaft

Torque
wrench

2012 © Delmar, Cengage Learning

5. Measure and adjust drive pinion depth.

The primary purpose of adjusting the pinion gear depth is to set the pinion and ring gear mesh. Pinion depth setting is the distance between a point—usually the gear end—of the pinion gear and the center line of the axles or differential case bearing bores. The pinion gear depth is normally adjusted by installing shims onto the pinion mounting. These measurements must be taken with the pinion bearings preloaded.

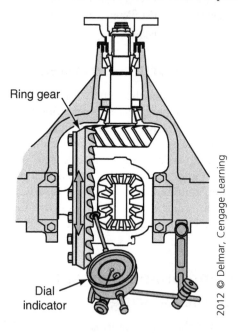

Ring gear

Dial indicator

2012 © Delmar, Cengage Learning

6. Measure and adjust drive pinion bearing preload (collapsible spacer or shim type).

When you replace ring and pinion sets, the first item to be installed and adjusted is the pinion gear. It has a bearing on the gear side and another on the companion flange side of it. The bearings must be preloaded to keep them from deflecting under the sideways loads the drive shaft and ring gear put on them. Preload is set either using shims between the bearings or a collapsible collar. Preload is measured in inch/pounds of resistance when turning the pinion nut. The pinion nut must never be loosened to obtain the specified turning torque. The pinion bearings should be lubricated when the turning torque is measured.

7. Measure and adjust differential (side) bearing preload and ring and pinion backlash (threaded adjuster or shim type).

Side bearing preload and backlash are adjusted by using either shims or threaded adjusters which press against the races of the side bearings. The processes for doing this vary by manufacturer and their instructions should be followed.

In all cases, the side bearing caps should be marked in relation to the case before removal, and the side bearings should be lubricated before installation.

8. Perform ring and pinion tooth contact pattern checks; determine needed adjustments.

Prussian Blue (a marking compound) is commonly used to check the tooth contact pattern on gear setup. A used gear set will have a shiny pattern on the gear teeth that can be visually inspected. Look for the pattern to be centered with either new or used gears after assembly. Carrier-bearing and pinion depth adjustments may be necessary to correct the depth and backlash of the gears. Do not concern yourself with the exact appearance of the pattern, as it varies by gear vendor. Take some time to look at the location of the pattern on the ring and pinion faces for the vehicle manufacturer or gear vendor.

2. Differential Case/Carrier Assembly (2 Questions)

1. Diagnose differential assembly noise and vibration problems; determine needed repairs.

Differentials perform their function only when a vehicle is turning a corner. Noises with any differential will usually be present only when turning unless there is a significant difference in tire dimensions from one side to the other. Grinding noises on turns are associated with differential side gears. Popping or crunching sounds are usually associated with differential cross-shaft failure and gear failure.

2. Remove and replace differential assembly.

On some types of axles, the pinion gear and differential assembly all come out of the axle housing as one assembly; in others, the bearing caps may be removed and just the differential carrier removed from the housing. In either case, the bearing caps should always be marked when removed to ensure they go back together correctly. The axle shafts must be removed before the differential assembly will come out. The shim packs and bearing races should be kept in order.

3. Inspect, measure, adjust, and replace differential pinion (spider) gears, shaft, side gears, thrust washers, side bearings, and case/carrier.

Use feeler gauges to check the clearance between differential side gears and thrust washers. Remove the differential side gears and pinions. Inspect all gears for abnormal wear, heat damage, chipped teeth, brinelling, and other wear. Inspect the differential pinion shaft for wear, cracks, lack of lubrication, and heat damage. When assembling differential gears and side gears, place the side gears and washers in the case. Then walk the pinions around the side gears until they align with the shaft hole; insert the shaft, spacer block, and lock pin. After installation, rotate the gears a few times and recheck thrust washer clearance with feeler gauges.

4. Measure differential case/carrier runout; determine needed repairs.

To check differential case runout, place the case in a set of V-blocks. Then place a dial indicator against the ring gear mounting flange and rotate the case to measure axial runout. Move the dial indicator pointer to the ring gear hub area of the case and rotate the case again to measure radial runout. If runout is out of limits in either direction, replace the case.

3. Limited Slip/Locking Differential (1 Question)

1. Diagnose limited slip differential noise, slippage, and chatter problems; determine needed repairs.

Differentials perform their function only when a vehicle is turning a corner. Noises with any differential will usually be present only when turning unless there is a significant difference in tire dimensions from one side to the other. Limited slip differentials are designed to lock the two wheels together via the differential when accelerating in a straight line. Open or non-limited slip differentials drive only one wheel during acceleration. Locking of the differential is accomplished through the use of clutches or spring-loaded plates that will release and slip during cornering. Noises, slippage, and chatter usually occur due to improper fluid in the axle or wear of the limited slip components. Special additives are used in most limited slip differentials to help provide smooth release and engagement of the differential components.

2. Inspect, drain, and refill with proper lubricant.

When you fill a transmission or differential case, fluid will come out of the fill hole as it reaches the full level. After draining differential fluid, inspect it for excessive metal particles. Silver- or steel-colored particles are signs of gear or bearing wear. Copper- or bronze-colored particles are signs of limited slip clutch disc wear.

3. Inspect, adjust, repair or replace limited slip or locking assembly components.

The friction plates have a minimum thickness specification and should be measured with a micrometer. The friction plate has to be removed from the clutch pack for this measurement. There is no way to measure a friction plate with a feeler gauge.

4. Axle Shafts and Housing (1 Question)

1. Diagnose rear axle shaft noise, vibration, and fluid leakage problems; determine needed repairs.

Noises in differentials can be from the axle bearings, ring and pinion, any of the differential bearings or differential side gears. Axle bearings usually can be isolated by the change in noise as the vehicle experiences different side loads or by raising the vehicle and running it in gear. Most axle bearing noises will subside dramatically when weight is taken off the wheels. Axle shaft bearings that are worn will cause the axle shaft to apply load on the axle shaft seal, and the seal will fail. Many axle shafts ride directly on the axle bearing and seem to be more susceptible to wear, which results in noise and fluid leaks.

2. Inspect and replace rear axle shaft wheel studs.

When you replace a wheel stud, a hammer can be used to carefully tap the old stud out, but when you install a new wheel stud, an installation tool should be used to avoid damage to the new stud. A torch should never be used to burn out a wheel stud because damage to the axle shaft and axle seal may occur.

3. Remove, inspect, adjust, and/or replace rear axle shafts, splines, seals, bearings, and retainers.

A worn axle shaft bearing can cause axle shaft seal failure. When the bearings wear out, axle side movement applies a greater load on the seal lip. Scored axle shafts in the seal area will damage the seal lip and cause the seal to fail. The axle shaft seal comes with a sprayed-on sealant and does not require any other sealant. The sealing lip of the axle shaft seal should be lubricated with a light coating of gear lube to prolong the seal life.

If the axle shaft is damaged near the seal area, the shaft should be discarded and replaced with a new one.

Splines must be inspected for wear and twisting.

4. Measure rear axle flange runout and shaft end-play; determine needed repairs.

To measure the axle shaft end-play, you will need to remove the wheel and tire assembly and the brake drum. A dial indicator is mounted or clamped to the axle housing or suspension. The axle shaft must be pushed into the housing all the way until it stops. Rest the dial indicator head on the face of the axle shaft flange. With the dial indicator set to zero, pull out on the axle shaft. The resulting dial indicator reading is the axle shaft end-play. The vehicle differential cover does not need to be removed.

Excessive runout could be caused by a bent axle shaft. A worn C-lock will cause excessive end-play. A worn bearing will cause fluid leakage. A bent housing is considered major damage and is often noticeable, and may be accompanied by chronic seal and axle bearing failure on that side of the housing.

5. Inspect axle housing and vent.

A damaged axle housing could cause fluid leakage at the damaged area. The housing should be thoroughly inspected when repairing an axel housing fluid leak. Another common problem for fluid leakage is the axle housing vent. If the vent gets plugged, the axle housing can build up excessive pressure, causing fluid to leak past gaskets and seals.

F. Four-Wheel Drive/All-Wheel Drive Component Diagnosis and Repair (8 Questions)

1. Diagnose drive assembly noise, vibration, shifting, leakage, and steering problems; determine needed repairs.

Worn U-joints may cause a squeaking or clunking noise and a vibration while driving straight ahead. Worn outer front-drive axle joints on a four-wheel drive (4WD) vehicle may cause a vibration while cornering.

When a vacuum-operated 4WD does not shift into 4WD, the engine vacuum may be low or the vacuum motor at the front differential may be damaged. Another cause could be bad or disconnected vacuum lines.

When attempting to isolate noises in 4WD systems, it is useful to determine if the noise is only present when operating in 4WD.

Correct tire size on all wheels is essential in 4WD/all-wheel drive (AWD) systems to prevent vibration, wind-up in the drive train, and shifting problems.

2. Inspect, adjust, and repair transfer case manual shifting mechanisms, bushings, mounts, levers, and brackets.

Specification measurements should be taken and recorded to aid in the installation of parts that have tolerances. All parts should be cleaned and lubricated before the assembly of the part. All components should be inspected for wear or damage.

The annulus gear is locked to the case so it cannot rotate. In 4WD low, the transfer case input shaft is driving the sun gear, which, in turn, is driving the planetary carrier. If the 4WD on the vehicle is not used often, the shift linkage could become rusted, and it will not work easily or at all. The shift linkage should be inspected for bushings that may be worn or deteriorated and need replacement. If the shift linkage does not move the linkage its full range, the transfer case may not operate in 4WD. The transfer case would still shift into gear if the front drive shaft universal joints are bound. Low fluid may cause damage to internal components but the transfer case will still engage. A manual shift transfer case has no electronic shift motor.

3. Remove and replace transfer case.

Transmission and transfer cases are removed as an assembly. Most newer 4WD vehicles have the transfer case bolted to the rear of the transmission. Not all transfer cases are made of cast iron; most new models have a lightweight aluminum case.

4. Disassemble transfer case; clean and inspect internal transfer case components; determine needed repairs.

Clean the transfer case with a steam cleaner, degreaser, or cleaning solvent. As you begin to disassemble the unit, pay close attention to the condition of its parts. Using a dial indicator, measure and record the end play of the input and mainshafts. This information will be needed during assembly of the unit to select the appropriate selective shims and washers. The transfer case parts should be cleaned with solvent, dried, and lubricated before assembly.

A plugged transfer case vent may cause seal leakage. A remote transfer case vent helps keep water out of the transfer case.

5. Reassemble transfer case.

If the drive chain in the transfer case is stretched and the drive sprocket teeth are worn, the chain could slip on the teeth and cause a loud clicking noise under acceleration.

It is important to follow manufacturer's procedures when reassembling the transfer case to avoid damaging shift forks and housings.

If the 4WD on the vehicle is not used often, the shift linkage could become rusted, and it will not work easily or at all. The shift linkage should be inspected for bushings that may be worn or deteriorated and need replacement. If the shift linkage does not move the linkage its full range, the transfer case may not operate in 4WD. The transfer case would still shift into gear if the front drive shaft universal joints are bound. Low fluid may cause damage to internal components but the transfer case will still engage. A manual shift transfer case has no electronic shift motor.

6. Check transfer case fluid level; drain and refill with proper fluid.

Oils break down over a period of time and miles. The transfer case oil should be drained and refilled with fresh oil at normal intervals. Most transfer cases require the fluid level be even with the bottom of the fill hole. Care should be taken to ensure that the proper type of lubricant is used. Some transfer cases use 80W-90 gear oil, some use ATF, and some use motor oil.

7. Inspect, service, and replace drive/propeller shaft and universal/CV joints.

When removing the front drive shaft, make sure the shaft is marked for proper reinstallation. Never allow a drive shaft to hang down with the weight on one joint; always support the disconnected end with a tie strap or bungee cord while removing the opposite end. Some CV shafts are secured using hex-headed bolts, while others use circlips. When inspecting front drive shafts and front CV joints, look for torn boots as well as any sign of looseness in the CV joints and/or the universal joints.

8. Inspect, service, and replace drive axle universal/CV joints and drive/half-shafts.

The drive axle CV joint should be inspected any time a torn boot is discovered. The CV joint and/or universal joint should be inspected for looseness, binding, and wear grooves. The inner end of most front-drive axles floats in the side gear while the outer end has a CV or universal joint, and is held in place by the wheel hub and spindle. When replacing, be sure to compare the old CV shaft with the new replacement to ensure the new shaft is an exact replacement. Some CV shafts require the installation of CV joint grease when replacing the shaft; these types of shafts usually bolt in place. Be sure to use the entire packet of grease, as this is a premeasured amount and all should be used.

9. Inspect, service, and replace wheel bearings, seals, and hubs.

Neither the automatic locking hubs nor the caps should be packed with grease. If they are packed with grease, they will not operate properly; the parts must move freely.

If only one bearing is bad, replace only that bearing. When a wheel bearing is replaced, the bearing race should also be replaced. The new bearing may fail if the race is not replaced.

10. Check transfer case and axle seals and all vents.

The only purpose of a remote vent is to keep moisture out of a differential assembly. This is needed on a 4WD vehicle in case the axle or transfer case is submerged in water.

11. Test, diagnose, and replace axle actuation and engagement systems (including: viscous, hydraulic, magnetic, mechanical, vacuum, and electrical/electronic).

Electronically-controlled actuation and engagements systems require special diagnostic test equipment. As a technician, you should have an understanding of electrical and electronic operational theory. Be sure to refer to manufacturer's recommended testing and service procedures. Damage to electronic components and control devices can result from improper testing and installation procedures. Some systems use a vacuum-actuated mechanical system. These types of engagement systems require the check of adequate vacuum to actuate the axle. To test this system, install a vacuum gauge to the hose going to the vacuum actuator and measure vacuum in two-wheel drive and four-wheel drive to determine correct operation.

12. Inspect tires for proper size and condition for vehicle application.

Tire size is crucial for proper transfer case differential operation. With mismatched tires, the transfer case differential or power transfer unit is constantly operating, causing premature failure due to excessive operation. The power transfer unit and transfer case differential are designed to allow the front and rear axles to operate at different speeds. However, mismatched tires can overwork these units.

Sample Preparation Exams

INTRODUCTION

Included in this section are a series of six individual preparation exams that you can use to help determine your overall readiness to successfully pass the Manual Drive Trains and Axles (A3) ASE certification exam. Located in Section 7 of this book you will find blank answer sheet forms you can use to designate your answers to each of the preparation exams. Using these blank forms will allow you to attempt each of the six individual exams multiple times without risk of viewing your prior responses.

Upon completion of each preparation exam, you can determine your exam score using the answer keys and explanations located in Section 6 of this book. Included in the explanation for each question is the specific task area being assessed by that individual question. This additional reference information may prove useful if you need to refer back to the task list located in Section 4 for additional support.

PREPARATION EXAM 1

1. A manual transmission grinds when attempting to select first or reverse gear only. Which of these could be the cause?

   A. A dry release bearing
   B. A chipped speed gear
   C. A warped clutch disc
   D. Insufficient clutch free travel

2. Technician A says broken transmission mount can cause excessive driveline angles. Technician B says an oil-soaked transmission mount should be inspected very closely for rubber deterioration and swelling. Who is correct?

   A. A only
   B. B only
   C. Both A and B
   D. Neither A nor B

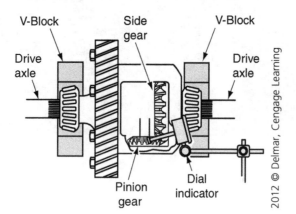

3. Which of these is the technician measuring in the illustration?
   A. Ring-gear backlash
   B. Differential end-play
   C. Side gear backlash
   D. Pinion gear backlash

4. A rear-wheel drive vehicle is being diagnosed for a speed-related vibration. Technician A says most joint working angles should not exceed 3 degrees. Technician B says two joint working angles on a common shaft should cancel each other out within 1 degree. Who is correct?
   A. A only
   B. B only
   C. Both A and B
   D. Neither A nor B

5. A transfer case is being removed for servicing. Technician A says the transfer case and transmission must be removed as a unit. Technician B says all yokes and flanges should be marked for correct reassembly. Who is correct?
   A. A only
   B. B only
   C. Both A and B
   D. Neither A nor B

6. A vehicle with limited slip differential is being diagnosed for chattering on a turn. Technician A says the differential may have the incorrect type of lubricant in it. Technician B says the last technician that serviced the differential may have not added the special limited-slip additive. Who is correct?
   A. A only
   B. B only
   C. Both A and B
   D. Neither A nor B

7. A vehicle with a manual transmission has been towed into the shop with a no-start complaint. The technician verifies the complaint and finds that the vehicle will not crank. Technician A says to check the clutch safety switch for an open circuit. Technician B says the clutch switch could be out of adjustment. Who is correct?
   A. A only
   B. B only
   C. Both A and B
   D. Neither A nor B

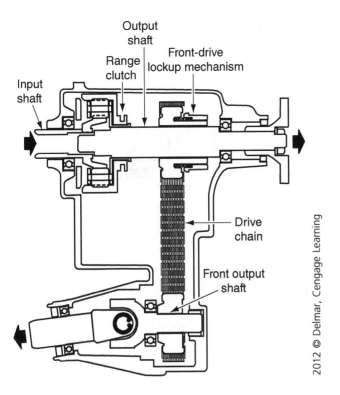

8. Technician A says if the drive chain shown in the illustration were to break, then the vehicle would still have power flow to the rear wheels. Technician B says low range can be achieved with a broken drive chain. Who is correct?

   A. A only

   B. B only

   C. Both A and B

   D. Neither A nor B

9. During a transmission overhaul, a technician finds a chipped tooth on the countershaft. Technician A says the countershaft and mating speed gear should be replaced. Technician B says only the damaged countershaft should be replaced. Who is correct?

   A. A only

   B. B only

   C. Both A and B

   D. Neither A nor B

10. A vehicle with four-wheel drive and tapered front-wheel bearing is being serviced. Technician A says the front wheel bearings should adjusted with 0.001–0.005 inches of end-play. Technician B says some late-model vehicles use unitized wheel bearings. Who is correct?

    A. A only

    B. B only

    C. Both A and B

    D. Neither A nor B

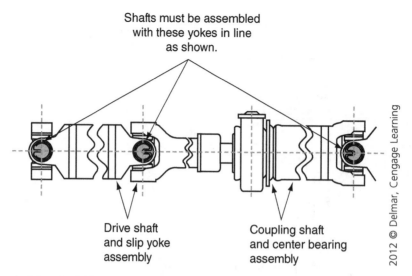

Shafts must be assembled
with these yokes in line
as shown.

Drive shaft
and slip yoke
assembly

Coupling shaft
and center bearing
assembly

2012 © Delmar, Cengage Learning

11. A drive shaft like the one shown in the illustration is reassembled after U-joint replacement without lining up the yokes. Technician A says this could cause driveline vibration. Technician B says the yokes must be in line for clearance. Who is correct?

    A. A only

    B. B only

    C. Both A and B

    D. Neither A nor B

12. A vehicle has a squealing noise coming from the clutch area when the clutch is disengaged. Technician A says the release bearing is probably dry and needs to be lubricated. Technician B says the damage has been done and the release bearing should be replaced. Who is correct?

    A. A only

    B. B only

    C. Both A and B

    D. Neither A nor B

13. A vehicle needs to have its pilot bearing replaced (bronze bushing type). Technician A says to make sure you check pilot-bearing bore runout. Technician B says it is necessary to lubricate the pilot bearing with high-quality white lithium grease. Who is correct?

    A. A only

    B. B only

    C. Both A and B

    D. Neither A nor B

14. Differential side-bearing preload is being adjusted. Technician A says that some RWD differentials require the selection of the correct shim thickness to adjust the side-bearing preload. Technician B says that some RWD differentials use threaded adjusters to set side-bearing preload. Who is correct?

    A. A only

    B. B only

    C. Both A and B

    D. Neither A nor B

15. A transaxle has had repeated drive axle seal leakage and replacement. Technician A says this problem may be caused by a plugged transaxle vent. Technician B says this problem may be caused by a worn outer-drive axle joint. Who is correct?

    A. A only

    B. B only

    C. Both A and B

    D. Neither A nor B

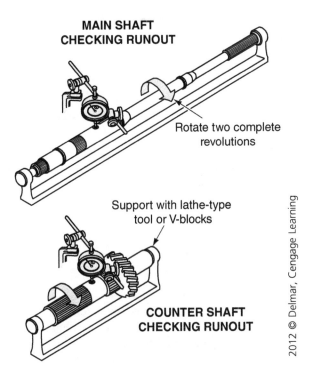

MAIN SHAFT CHECKING RUNOUT

Rotate two complete revolutions

Support with lathe-type tool or V-blocks

2012 © Delmar, Cengage Learning

COUNTER SHAFT CHECKING RUNOUT

16. Refer to the supplied illustration. Technician A says if the mainshaft runout is above specifications, then the mainshaft must be replaced. Technician B says the mainshaft can be straightened using an acetylene torch and hydraulic press. Who is correct?

    A. A only

    B. B only

    C. Both A and B

    D. Neither A nor B

17. A limited-slip differential with clutch packs is being overhauled. Technician A says the clutch pack discs should be installed alternately (steel, fiber, steel). Technician B says the clutch pack discs should be installed with the steel discs on the right and fiber discs on the left. Who is correct?

    A. A only

    B. B only

    C. Both A and B

    D. Neither A nor B

18. In a hydraulic clutch system, the clutch fails to disengage properly when the clutch pedal is fully depressed. The cause of this problem might be which of the following?

    A. Less than specified clutch pedal freeplay

    B. Air in the clutch hydraulic system

    C. Worn clutch facings

    D. A scored pressure plate

19. A truck with a 4-speed transmission has a growling noise coming from the transmission in all gears except fourth. In fourth, the noise is almost completely gone. Technician A says the input shaft bearing could cause the noise. Technician B says the countershaft bearing could be causing the noise. Who is correct?

    A. A only

    B. B only

    C. Both A and B

    D. Neither A nor B

20. A transaxle has been disassembled for intermediate shaft bearing replacement. Technician A says snap rings should not be reused and that new ones should be installed. Technician B says bearings that pass inspection can be reused as long as they were removed using a puller and not driven off with a hammer. Who is correct?

    A. A only

    B. B only

    C. Both A and B

    D. Neither A nor B

21. A vehicle with automatic locking hubs is being diagnosed for slow engagement. Technician A says that many vehicles with automatic locking hubs require as much as a full tire rotation in a specified direction in order for the hub to engage and disengage. Technician B says the driver does not have to leave the vehicle to engage the automatic locking hubs. Who is correct?

    A. A only

    B. B only

    C. Both A and B

    D. Neither A nor B

22. All of the following statements are true about removing and installing axle shafts EXCEPT:

    A. Inspect the shaft near the seal area and repair with fine sand paper, as needed.

    B. The axle shaft seals should be replaced, not reused.

    C. The axle shaft should be stood straight up in a vertical position when removed.

    D. Some axles require a slide hammer to remove them.

23. The second-speed gear clutch teeth and blocking ring teeth are badly worn on a transaxle. This problem may cause which of the following conditions?

    A. A growling noise while driving in second gear

    B. A vibration while accelerating in second gear

    C. Hard shifting in second and third gear

    D. The transaxle to jump out of second gear

24. An extension housing has burrs and gouges on the mating surface. Technician A says that if they are not excessive, they can be repaired with a file. Technician B says that the mating surfaces are machined surfaces and they should only be repaired with fine grit sandpaper. Who is correct?

    A. A only

    B. B only

    C. Both A and B

    D. Neither A nor B

25. A remote vent on a differential on a 4WD vehicle is used to:

    A. Increase pressure in the differential

    B. Keep moisture out of the differential

    C. Keep lubricant from coming out of the differential

    D. Add lubricant to the differential

26. Technician A says that if the differential is overfilled, then it will cause axle seals to leak. Technician B says that worn axle bearings will cause axle seals to leak. Who is correct?

    A. A only

    B. B only

    C. Both A and B

    D. Neither A nor B

27. A new universal joint has been installed in a vehicle. Technician A says the universal joint should be greased until grease purges out of all four caps. Technician B says that the universal joint is greased with a good quality lithium-based grease meeting *NLGI (national grease lubricating institute) Grade 1 or Grade 2.* Who is correct?

    A. A only

    B. B only

    C. Both A and B

    D. Neither A nor B

28. Front universal joints are being replaced. Technician A says the new U-joints will still need to be greased after assembly. Technician B says to pay close attention to the location of the zerk fitting—if installed incorrectly, then the zerks cannot be reached with a grease gun. Who is correct?

    A. A only

    B. B only

    C. Both A and B

    D. Neither A nor B

29. Fluid is being changed in a transaxle. Technician A says the proper fluid level is one finger joint through the fill hole. Technician B says most transaxles use 90-weight gear lube. Who is correct?

    A. A only

    B. B only

    C. Both A and B

    D. Neither A nor B

30. All of the following statements about differential case/carrier and ring-gear removal and replacement are true EXCEPT:

    A. The ring-gear runout should be measured before removal of the case/carrier and ring-gear assembly.

    B. The case/carrier side-play should be measured before removal of the case/carrier and ring-gear assembly.

    C. The side bearing caps should be marked in relation to the housing before removal of the case/carrier and ring-gear assembly.

    D. The side bearing should be clean and dry before installation of the case/carrier and ring-gear assembly.

31. A transmission is being disassembled for overhaul. Technician A says care should be taken to thoroughly inspect each component for damage as it is disassembled. Technician B says some transmission components have index marks that must be aligned when reassembled. Who is correct?

    A. A only

    B. B only

    C. Both A and B

    D. Neither A nor B

32. When adjusting the pinion depth on the ring gear, technician A says to replace the collapsible spacer. Technician B says that pinion depth can also be adjusted by installing a selective pinion bearing race. Who is correct?

    A. A only

    B. B only

    C. Both A and B

    D. Neither A nor B

33. A transaxle has fluid leaking from the output shaft seal on an all-wheel drive vehicle. The technician removes the seal for inspection and cannot see any problem with the seal. Technician A says to check for a damaged yoke seal surface. Technician B says the vent could be plugged. Who is correct?

    A. A only

    B. B only

    C. Both A and B

    D. Neither A nor B

34. A four-wheel drive vehicle has a jerking feel in the steering wheel when turning sharply with the four-wheel drive engaged. Technician A says this could be normal for many four-wheel drive vehicles. Technician B says if the vehicle has CV joints in the front axle and a center differential, then this would not be normal. Who is correct?

    A. A only

    B. B only

    C. Both A and B

    D. Neither A nor B

35. In a manual 5-speed transmission, the reverse idler gear is in mesh with what other gear?

    A. First speed gear
    B. Fifth speed gear
    C. Main drive gear
    D. Reverse speed gear

36. Technician A says an unbalanced drive shaft causes a vibration that increases as vehicle speed increases. Technician B says an unbalanced drive shaft causes a vibration that decreases as vehicle speed increases. Who is correct?

    A. A only
    B. B only
    C. Both A and B
    D. Neither A nor B

37. A vehicle has an inoperative speedometer. Technician A says the vehicle speed sensor could be defective. Technician B says the speedometer drive gear could be broken. Who is correct?

    A. A only
    B. B only
    C. Both A and B
    D. Neither A nor B

38. Technician A says vehicle speed sensors can be tested with an ammeter. Technician B says when you check a vehicle speed sensor signal, you should use an oscilloscope while the vehicle is being driven. Who is correct?

    A. A only
    B. B only
    C. Both A and B
    D. Neither A nor B

39. A vacuum-shifted 4WD system does not shift into 4WD. Technician A says the vacuum hose that supplies the 4WD vacuum motor may be cracked. Technician B says the vacuum motor at the front axle may be the problem. Who is right?

    A. A only
    B. B only
    C. Both A and B
    D. Neither A nor B

40. A vehicle is being diagnosed for a driveline vibration. Technician A says the drive shaft should be checked for proper phasing. Technician B says that drive shaft runout should be checked. Who is correct?

    A. A only
    B. B only
    C. Both A and B
    D. Neither A nor B

PREPARATION EXAM 2

1. Technician A says a close inspection of the transmission input shaft is part of a clutch replacement procedure. Technician B says the transmission input shaft can be replaced without completely disassembling the transmission on some transmissions. Who is correct?

   A. A only
   B. B only
   C. Both A and B
   D. Neither A nor B

2. Ring-gear runout is being measured on a differential ring gear. Technician A says if ring-gear runout is excessive, then the ring gear can be moved 180 degrees on the ring-gear carrier and rechecked. Technician B says ring-gear runout should be checked before disassembly. Who is correct?

   A. A only
   B. B only
   C. Both A and B
   D. Neither A nor B

3. A 5-speed transmission has been rebuilt because of a chipped tooth. The transmission now gets hung in two gears. Technician A says too stout a detent spring was installed. Technician B says the interlock should be inspected for proper operation. Who is correct?

   A. A only
   B. B only
   C. Both A and B
   D. Neither A nor B

4. Technician A says the first step in transaxle removal is to disconnect the negative battery cable from the battery. Technician B says guide pins should be used to help support the transmission as it is slid out. Who is correct?

   A. A only
   B. B only
   C. Both A and B
   D. Neither A nor B

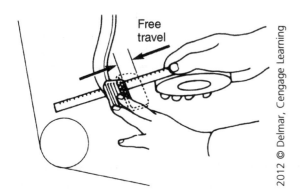

5. Technician A says the clutch safety switch free travel is being measured in the illustration. Technician B says that as the clutch disc wears, this free travel will decrease. Who is correct?

   A. A only
   B. B only
   C. Both A and B
   D. Neither A nor B

6. Technician A says that the removal of a drive shaft usually requires partial disassembly of the front suspension. Technician B says the first step in drive shaft removal is to loosen the rear differential yoke nut. Who is correct?

   A. A only
   B. B only
   C. Both A and B
   D. Neither A nor B

7. Technician A says the collapsible pinion shaft spacer may be reused if the differential is disassembled and overhauled. Technician B says that after proper pinion bearing preload is set, pinion depth is adjusted by backing off the pinion flange nut. Who is right?

   A. A only
   B. B only
   C. Both A and B
   D. Neither A nor B

8. An all-wheel drive vehicle has a vibration that is more noticeable while cornering. Technician A says the U-joints in the drive shaft may be worn. Technician B says the outboard front axle joints may be worn. Who is right?

   A. A only
   B. B only
   C. Both A and B
   D. Neither A nor B

9. The counter/cluster gear shaft and needle bearings in a 4-speed transmission are pitted and scored. Technician A says the transmission may have a growling noise with the engine idling, transmission in neutral, and the clutch pedal out. Technician B says the transmission may have a growling noise while driving in any gear except fourth gear. Who is right?

    A. A only
    B. B only
    C. Both A and B
    D. Neither A nor B

10. A front-wheel drive vehicle has a clicking noise while turning. Technician A says this could be caused by a worn outer-drive axle joint. Technician B says this could be caused by a worn front wheel bearing. Who is right?

    A. A only
    B. B only
    C. Both A and B
    D. Neither A nor B

11. A rear-wheel (RWD) drive vehicle has a clunking noise when accelerating from a stop. Technician A says if this vehicle uses a center-support bearing, then it should be inspected for proper operation. Technician B says the center-support bearing can be shimmed to adjust driveline angle. Who is correct?

    A. A only
    B. B only
    C. Both A and B
    D. Neither A nor B

12. A transaxle is being diagnosed for hard shifting. Technician A says the lubricant should be checked for proper viscosity. Technician B says if a highly viscous oil is used, then the hard shifting will be more noticeable in colder weather. Who is correct?

    A. A only
    B. B only
    C. Both A and B
    D. Neither A nor B

13. A four-wheel drive with a manual shifting linkage is being adjusted. Technician A says some transfer cases require the 4WD lever be placed in a specific position before adjustments can be made. Technician B says some transfer cases require the use of a spacer of a certain size to position the lever properly for adjustment. Who is correct?

    A. A only
    B. B only
    C. Both A and B
    D. Neither A nor B

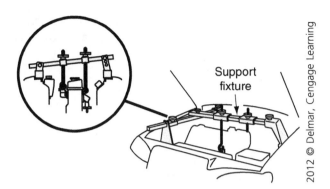

14. Technician A says the transaxle shown in the illustration is being removed and the support fixture is used to support the engine during transaxle removal. Technician B says the support fixture is used to help position the engine during transaxle installation. Who is correct?

  A. A only
  B. B only
  C. Both A and B
  D. Neither A nor B

15. During a clutch replacement job, the flywheel is found to have too much runout. Technician A says sometimes the flywheel runout can be adjusted by re-torquing the flywheel. Technician B says excessive flywheel runout can cause poor release symptoms. Who is correct?

  A. A only
  B. B only
  C. Both A and B
  D. Neither A nor B

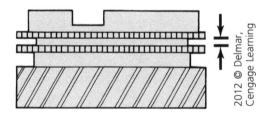

16. In reference to the illustration, Technician A says that as the synchronizer blocker ring wears, the distance between the two arrows will increase. Technician B says this distance is measured with a feeler gauge. Who is correct?

  A. A only
  B. B only
  C. Both A and B
  D. Neither A nor B

17. A four-wheel drive vehicle has had a premature failure of the viscous coupler. Technician A says the tires should be checked for correct brand type and size. Technician B says that low pressure, as little as 5 psi, can effect operation of the viscous coupler. Who is correct?

   A.   A only

   B.   B only

   C.   Both A and B

   D.   Neither A nor B

18.   Technician A says a plugged transfer case vent can cause fluid leakage from a seal. Technician B says that a remotely located transfer-case vent helps to prevent water from entering the transfer case when driving through high waters. Who is right?

   A.   A only

   B.   B only

   C.   Both A and B

   D.   Neither A nor B

19. A rear-wheel drive vehicle has a vibration that increases in relation to vehicle speed. Technician A says the counter weight may have fallen off the drive shaft. Technician B says some of the wheels and tires may be out of balance. Who is right?

   A.   A only

   B.   B only

   C.   Both A and B

   D.   Neither A nor B

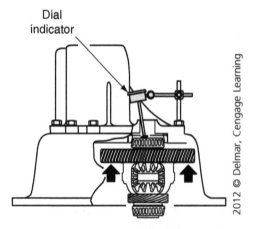

Dial indicator

2012 © Delmar, Cengage Learning

20.   The measurement in the illustration determines the proper:

   A.   Bearing wear

   B.   Side bearing preload

   C.   Side gear end-play

   D.   Bearing race depth

21. Technician A says the primary difference between full-time four-wheel drive and part-time four-wheel drive is the center differential. Technician B says the four-wheel drive on a vehicle with part-time four-wheel drive is for off-road only. Who is correct?

    A. A only
    B. B only
    C. Both A and B
    D. Neither A nor B

22. A vehicle has an oil-contaminated clutch disc. Technician A says the source of the oil could be the rear main seal in the engine. Technician B says the source of the oil could be from the engine valve covers. Who is correct?

    A. A only
    B. B only
    C. Both A and B
    D. Neither A nor B

23. Technician A says misadjusted shift linkage can cause the transmission to jump out of gear. Technician B says misadjusted shift linkage can cause the transmission to have hard shifting. Who is correct?

    A. A only
    B. B only
    C. Both A and B
    D. Neither A nor B

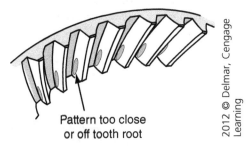

Pattern too close
or off tooth root

2012 © Delmar, Cengage Learning

24. Technician A says the pinion depth shown in the illustration is too deep. Technician B says ring-gear backlash should be decreased. Who is correct?

    A. A only
    B. B only
    C. Both A and B
    D. Neither A nor B

25. A transaxle is being disassembled for overhaul. Technician A says the input and main shaft end-play should be measured before disassembly. Technician B says the input shaft should be rotated and the transaxle shifted through the gears including reverse gear. Who is correct?

    A. A only
    B. B only
    C. Both A and B
    D. Neither A nor B

26. Technician A says the flywheel can be shimmed if the crankshaft end-play is found to be excessive. Technician B says the flywheel runout should be checked every time the clutch is replaced. Who is correct?

    A. A only

    B. B only

    C. Both A and B

    D. Neither A nor B

27. Technician A says a limited slip differential can be checked using a torque wrench. Technician B says a limited slip differential can be checked using a spring scale. Who is correct?

    A. A only

    B. B only

    C. Both A and B

    D. Neither A nor B

28. An axle shaft on a rear-wheel drive has excessive end-play. Technician A says the problem could be a faulty axle bearing. Technician B says the problem could be a worn C-lock retainer. Who is correct?

    A. A only

    B. B only

    C. Both A and B

    D. Neither A nor B

29. Technician A says the tire speed index has no effect on all-wheel drive operation. Technician B says mixing tire brands can have an effect on all-wheel drive vehicles. Who is correct?

    A. A only

    B. B only

    C. Both A and B

    D. Neither A nor B

30. Technician A says it is acceptable to reuse lock washers when overhauling a transmission. Technician B says anaerobic sealers cure in the presence of air. Who is correct?

    A. A only

    B. B only

    C. Both A and B

    D. Neither A nor B

31. Technician A says guide studs should be used when removing a transmission. Technician B says guide studs should be used when installing a transmission. Who is correct?

    A. A only

    B. B only

    C. Both A and B

    D. Neither A nor B

32. Technician A says if a spreader is being used to remove a differential from the housing assembly, then the housing should not be spread more than 0.030 inches (0.762 mm). Technician B says if a spreader is not available, then the differential can be pried out using a pry bar and a piece of wood. Who is correct?

    A. A only

    B. B only

    C. Both A and B

    D. Neither A nor B

33. Technician A says if the transfer-case shaft assembly end-play is excessive, then selective snap rings may be used to get the end-play within tolerances. Technician B says some transfer cases use shims to adjust transfer-case shaft assembly end-play. Who is correct?

    A. A only

    B. B only

    C. Both A and B

    D. Neither A nor B

34. A speedometer is inoperative on a front-wheel drive vehicle. Technician A says the vehicle speed sensor could be the problem and should be checked. Technician B says a scan tool can be used to verify that the vehicle speed sensor (VSS) is operating correctly. Who is correct?

    A. A only

    B. B only

    C. Both A and B

    D. Neither A nor B

35. Technician A says the drive shaft should be marked before removing it during transmission removal. Technician B says some transmissions must be removed as a unit with the engine. Who is correct?

    A. A only

    B. B only

    C. Both A and B

    D. Neither A nor B

36. A front-wheel drive vehicle has a vibration in the steering wheel at highway speeds. Technician A says the inner CV joint could be worn and not operating smoothly. Technician B says the front wheels may be out of balance. Who is correct?

    A. A only

    B. B only

    C. Both A and B

    D. Neither A nor B

37. A rear-wheel drive vehicle has a growling noise coming from the rear axle that is only noticed in turns. Technician A says the problem is in the differential. Technician B says it could be worn carrier bearings. Who is correct?

    A. A only

    B. B only

    C. Both A and B

    D. Neither A nor B

38. Technician A says the transaxle must be disassembled in order to replace the differential on a FWD vehicle. Technician B says most transaxles are lubricated by gear rotation and troughs and funnels. Who is correct?

    A. A only
    B. B only
    C. Both A and B
    D. Neither A nor B

39. Technician A says transfer cases are lubricated with automatic transmission fluid (ATF). Technician B says transfer cases are lubricated with gear oil. Who is correct?

    A. A only
    B. B only
    C. Both A and B
    D. Neither A nor B

40. Technician A says the side-bearing preload on a transaxle can be set using threaded adjusters. Technician B says the side-bearing preload on a transaxle can be set using shims. Who is correct?

    A. A only
    B. B only
    C. Both A and B
    D. Neither A nor B

PREPARATION EXAM 3

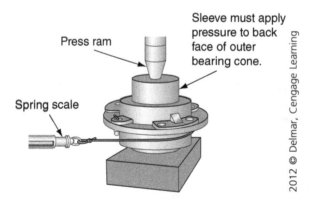

1. Technician A says pinion bearing preload is being set in the illustration. Technician B says the press simulates the clamping force of the yoke without having to actually install the yoke. Who is correct?

    A. A only
    B. B only
    C. Both A and B
    D. Neither A nor B

2. Clutch housing bore runout is excessive on a RWD vehicle. Technician A says the symptom for excessive clutch housing bore runout is premature pilot bearing/bushing wear. Technician B says the maximum general specification for clutch housing bore runout is 0.020 inches (0.50 mm). Who is correct?

    A. A only
    B. B only
    C. Both A and B
    D. Neither A nor B

3. Technician A says the idler gear is used for reverse gear only. Technician B says the idler gear end-play is controlled by thrust washers. Who is correct?

    A. A only
    B. B only
    C. Both A and B
    D. Neither A nor B

4. Technician A says when checking the transmission fluid always ensure the vehicle is level. Technician B says changing the fluid is important to remove metal contaminants that can cause an increase of wear on internal transmission components. Who is correct?

    A. A only
    B. B only
    C. Both A and B
    D. Neither A nor B

5. Technician A says that drive shaft runout should be measured 1 inch from the seam of the yoke. Technician B says the drive shaft runout should be measured at the center of the drive shaft. Who is correct?

    A. A only
    B. B only
    C. Both A and B
    D. Neither A nor B

6. All of the following are correct procedures when servicing wheel bearing EXCEPT:

    A. Ensure an end-play of 0.001–0.005 inches (0.0254–0.127 mm).
    B. The wheel bearing is cleaned in solvent and repacked.
    C. A new grease seal is installed.
    D. Use compressed air to spin-dry the bearing.

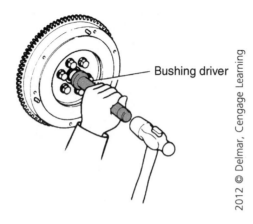

Bushing driver

2012 © Delmar, Cengage Learning

7. What is the technician doing in the above illustration?
    A. Installing a pilot bearing
    B. Removing a pilot bearing
    C. Installing a release bearing
    D. Removing a release bearing

8. A transaxle is jumping out of third gear when driving on a rough terrain. Technician A says the detent spring could be broken or weak. Technician B says the detent could be worn. Who is correct?
    A. A only
    B. B only
    C. Both A and B
    D. Neither A nor B

9. A drive axle has repeated pinion seal failure due to leaks. Technician A says the sealing surface of the seal should be inspected for a wear groove. Technician B says the collapsible spacer may need replacing to fix the seal failure complaint. Who is correct?
    A. A only
    B. B only
    C. Both A and B
    D. Neither A nor B

10. Excessive input shaft end-play in a 4-speed transmission may cause the transmission to jump out of:
    A. First gear
    B. Second gear
    C. Fourth gear
    D. Reverse gear

11. A front-wheel drive car has a clicking noise in the front while turning. Technician A says this could be caused by a worn outer-drive axle joint. Technician B says this could be caused by a front wheel bearing. Who is correct?
    A. A only
    B. B only
    C. Both A and B
    D. Neither A nor B

12. All of the following are true about transfer case disassembly and inspection and assembly EXCEPT:

    A. The transfer case chain should be inspected for stretching and looseness.

    B. Thrust washer thickness should be measured to check for wear or for proper select fit.

    C. Specification measurements should have been taken during disassembly.

    D. All parts should be cleaned before installation and assembled dry.

13. Technician A says the marcel spring allows smooth clutch engagement and longer drive train life. Technician B says the dampening springs compress during clutch engagement. Who is correct?

    A. A only

    B. B only

    C. Both A and b

    D. Neither A nor B

14. Technician A says if the transmission input shaft shows wear, then only on the side of the spines it can be reused. Technician B says the front pilot area of the input shaft should be inspected. Who is correct?

    A. A only

    B. B only

    C. Both A and B

    D. Neither A nor B

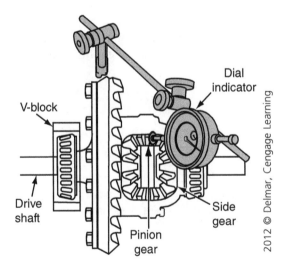

15. Technician A says if the measurement being taken in the illustration is excessive, then the pinion gear thrust washer must be replaced. Technician B says this measurement can determine gear wear. Who is correct?

    A. A only

    B. B only

    C. Both A and B

    D. Neither A nor B

16. Technician A says a worn U-joint may cause a squeaking noise. Technician B says a heavy vibration that only occurs during acceleration may be caused by a worn, centering ball and socket on a double-Cardan U-joint. Who is right?

    A. A only

    B. B only

    C. Both A and B

    D. Neither A nor B

17. Technician A says in order to measure differential carrier runout the ring gear should be removed. Technician B says the side bearing caps must be installed. Who is correct?

    A. A only

    B. B only

    C. Both A and B

    D. Neither A nor B

18. The electric shift motor fails to operate on a transfer case. Technician A says the voltage should be checked at the shift motor electrical connector; if the voltage is equal to source voltage, then the shift motor needs replacing. Technician B says the power and ground sides of the shift motor should be tested for voltage drop before any component is replaced. Who is correct?

    A. A only

    B. B only

    C. Both A and B

    D. Neither A nor B

19. Technician A says if clutch housing runout is out of limits, then shims can be used to correct the runout. Technician B says if shims are not available they can be cut from shim stock. Who is correct?

    A. A only

    B. B. only

    C. Both A and B

    D. Neither A nor B

20. A front-wheel drive vehicle has a clunking noise during acceleration or deceleration. Technician A says if the noise gets louder on a turn, then the problem is an inboard CV joint. Technician B says the problem is a worn-out inboard CV joint. Who is correct?

    A. A only

    B. B only

    C. Both A and B

    D. Neither A nor B

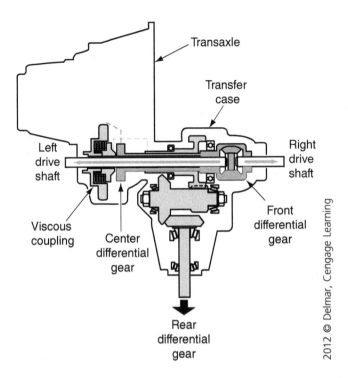

21. In reference to the illustration, technician A says if the center differential gear failed, then there would be no power flow to the rear wheels. Technician B says if the viscous coupling fails, then it must be replaced as a unit. Who is correct?

    A. A only

    B. B only

    C. Both A and B

    D. Neither A nor B

22. A transmission has a growling noise in neutral, with the clutch engaged. Technician A says the output shaft bearing could be the problem. Technician B says the input shaft bearing could be the problem. Who is correct?

    A. A only

    B. B only

    C. Both A and B

    D. Neither A nor B

23. Technician A says a misaligned engine and transaxle cradle may cause driveline vibrations. Technician B says a misaligned engine and transaxle cradle may affect front suspension angles. Who is right?

    A. A only

    B. B only

    C. Both A and B

    D. Neither A nor B

24. To measure drive shaft runout, you should do all of the following EXCEPT:

    A. Place the vehicle transmission in neutral.

    B. Use a magnetic-base dial indicator.

    C. Roll the drive shaft on a flat surface to check for damage.

    D. Clean the drive shaft surface for an accurate runout check.

25. Technician A says the proper level of lubricant on a limited-slip differential is one finger joint in the fill hole. Technician B says some limited-slip differentials require a special oil additive. Who is correct?

    A. A only

    B. B only

    C. Both A and B

    D. Neither A nor B

26. Excessive noise coming from the transfer case may be caused by all of the following EXCEPT:

    A. Low fluid level

    B. Misalignment of the transfer-case drive chain

    C. A worn universal joint

    D. A damaged output-shaft bearing

27. A transaxle has brass flakes in its oil during an oil change. Technician A says this is a sign that the bearings are beginning to chip and flake. Technician B says the synchronizers are beginning to flake and wear. Who is correct?

    A. A only

    B. B only

    C. Both A and B

    D. Neither A nor B

28. Technician A says that if a clutch noise only happens when the pedal is all of the way to the floor, then the release bearing is probably the cause. Technician B says the noise that happens only when the engine is idling in neutral with the clutch engaged is probably due to an input-shaft bearing. Who is correct?

    A. A only

    B. B only

    C. Both A and B

    D. Neither A nor B

29. A vehicle has a growling noise from the rear end that gets louder on a sharp right turn. Technician A says the vehicle has an axle bearing beginning to fail. Technician B says the differential side bearing is causing the noise. Who is correct?

    A. A only

    B. B only

    C. Both A and B

    D. Neither A nor B

30. Technician A says the hydraulic clutch system can be bled by putting fluid into the slave cylinder bleed screw. Technician B says that a hydraulic clutch system can be bled by letting fluid out of the slave cylinder bleed screw. Who is correct?

    A. A only
    B. B only
    C. Both A and B
    D. Neither A nor B

31. A vehicle has a vibration above 30 mph in the rear. Technician A says the driveline and companion flanges should be checked for excessive runout. Technician B says to disconnect the flanges and reconnect them 180 degrees apart if runout is found. Who is correct?

    A. A only
    B. B only
    C. Both A and B
    D. Neither A nor B

32. Technician A says the cluster gears are a major component of the transmission lubrication system. Technician B says some transmissions use oiling troughs to guide the oil to critical areas. Who is correct?

    A. A only
    B. B only
    C. Both A and B
    D. Neither A nor B

33. All of the following are part of the rear oil-seal replacement procedure on a transmission EXCEPT:

    A. Remove the drive shaft.
    B. Lubricate the lip of the new seal.
    C. Drive a new seal in place with a hammer and brass punch.
    D. Torque universal joint cap screws.

34. The transfer clutch in an all-wheel drive vehicle takes the place of which one of the following?

    A. Reduction gears
    B. Torque converter
    C. Interaxle differential
    D. Drive chain

35. A rear-wheel drive vehicle is being diagnosed for axle noise. Technician A says grinding noise only heard when going around a curve or turning is probably in the one of the axle bearings. Technician B says a clicking noise only heard when going around a curve or turning is probably in the outer constant-velocity joint. Who is correct?

    A. A only
    B. B only
    C. Both A and B
    D. Neither A nor B

36. When checking a diagnostic trouble code (DTC) for the output shaft speed sensor, which of the following should NOT be done next after reading the code?

    A. Install a new sensor.

    B. Inspect the harness for continuity.

    C. Check the service manual.

    D. Inspect the connector.

37. A 4-speed transaxle makes a clunking noise when driven in first gear and reverse. Technician A says that the gear on the countershaft could be the problem. Technician B says that the reverse idler gear could be the problem. Who is right?

    A. A only

    B. B only

    C. Both A and B

    D. Neither A nor B

38. An electronic transfer case does not engage. Which of the following would be the LEAST LIKELY cause?

    A. A bad electric shift motor

    B. A blown fuse

    C. The 4WD engage switch

    D. A rusted linkage

39. The best way to check mating surfaces for warpage on a transaxle case is to use which of the following?

    A. Straight edge

    B. Dial indicator

    C. Micrometer

    D. Flat surface

40. Technician A says that when a wheel bearing is replaced, the bearing race should also be replaced. Technician B says if the wheel bearing is a unitized bearing, then the race does not need replacing. Who is right?

    A. A only

    B. B only

    C. Both A and B

    D. Neither A nor B

PREPARATION EXAM 4

1. Technician A says to put the transmission in low gear to test for a slipping clutch. Technician B says that as a mechanical linkage clutch disc wears, the free travel becomes less. Who is correct?

    A. A only
    B. B only
    C. Both A and B
    D. Neither A nor B

2. Technician A says that repeated extension-housing seal failures may be caused by a worn extension-housing bushing. Technician B says that a worn countershaft bearing may cause an extension-housing bushing to wear prematurely. Who is correct?

    A. A only
    B. B only
    C. Both A and B
    D. Neither A nor B

3. A transaxle has had repeated drive-axle seal leakage and replacements. Technician A says this problem may be caused by a plugged transaxle vent. Technician B says this problem may be caused by a worn outer constant-velocity joint. Who is correct?

    A. A only
    B. B only
    C. Both A and B
    D. Neither A nor B

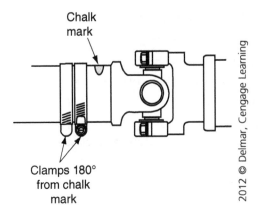

4. Technician A says the hose clamps in the illustration are for drive shaft balance. Technician B says the chalk mark shows the runout location. Who is correct?

    A. A only
    B. B only
    C. Both A and B
    D. Neither A nor B

5. All of the following statements about a limited-slip differential are true EXCEPT:

    A. Fiber plates are splined to the side gear.

    B. A special lubricant additive is needed.

    C. Each clutch pack uses a preload spring.

    D. The breakaway torque is checked with an inch pound torque wrench.

6. All of the following statements regarding 4WD front-drive axles and joints are true EXCEPT:

    A. The joint should be coated with good marine-grade NLG 2 grease.

    B. A worn CV joint may cause a clicking noise when cornering.

    C. The inner tripod joint is held on the axle shaft with a cir-clip.

    D. A special tool is required to install the boot clamp.

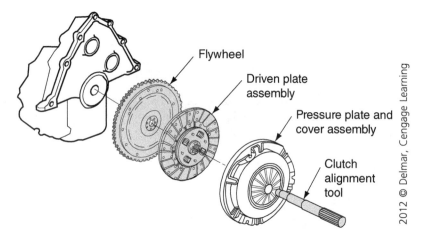

7. All of the following are true about the clutch alignment tool EXCEPT:

    A. Align the pilot bearing with the clutch hub opening.

    B. Align the pressure plate with the flywheel.

    C. Hold the clutch disc in alignment during pressure plate installation.

    D. Make one out of an old transmission input shaft.

8. Technician A says that a back-up lamp switch is a normally open switch. Technician B says that a back-up lamp switch can have power to the switch in Run and is closed when the vehicle is shifted into reverse. Who is correct?

    A. A only

    B. B only

    C. Both A and B

    D. Neither A nor B

9. Technician A says vehicle speed sensors can be tested with an ammeter. Technician B says when you check a vehicle speed sensor signal, you should use an oscilloscope while the vehicle is being driven or run on the rack. Who is correct?

    A. A only
    B. B only
    C. Both A and B
    D. Neither A nor B

10. Technician A says the dial indicator should be positioned near the rear of the drive shaft to measure drive shaft runout. Technician B says if the drive shaft runout is out of limits, then the drive shaft may be straightened using a hydraulic press and straight edge. Who is correct?

    A. A only
    B. B only
    C. Both A and B
    D. Neither A nor B

11. Technician A says that excessive pinion-nut torque may cause a clunking noise during acceleration or deceleration. Technician B says that excessive pinion-nut torque may cause a growling noise with the vehicle in motion. Who is correct?

    A. A only
    B. B only
    C. Both A and B
    D. Neither A nor B

12. When a torque management transfer case is operating in automatic four-wheel drive (A4WD) and the front drive shaft begins to turn faster than the rear drive shaft, what happens?

    A. The computer energizes the viscous coupling.
    B. The computer reduces the throttle opening.
    C. The computer increases the clutch coil duty-cycle.
    D. The computer reduces the torque to the rear wheels.

13. Technician A says if too much material is machined away during flywheel resurfacing, then the torsion springs on the clutch disc may contact the flywheel bolts, resulting in noise while engaging and disengaging. Technician B says if excessive material is removed when the flywheel is resurfaced, then the slave cylinder rod may not have enough travel to release the clutch properly. Who is correct?

    A. A only
    B. B only
    C. Both A and B
    D. Neither A nor B

14. The shift lever adjustment on a transmission with external linkage is usually performed with the transmission in which gear?

    A. First
    B. Second
    C. Neutral
    D. Reverse

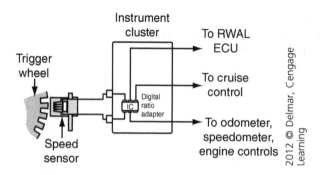

15. Technician A says if the speedometer shown does not work on a vehicle with a transaxle, then check the signal at the vehicle speed sensor. Technician B says the trigger wheel could be defective. Who is correct?

    A. A only

    B. B only

    C. Both A and B

    D. Neither A nor B

16. A rear-wheel drive vehicle has a vibration that increases in relation to vehicle speed. Technician A says the balance weight may have fallen off the drive shaft. Technician B says it might have a wheel out of balance. Who is correct?

    A. A only

    B. B only

    C. Both A and B

    D. Neither A nor B

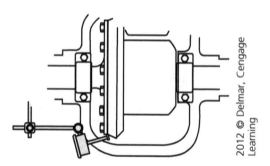

17. Excessive ring-gear runout on the dial indicator shown in the illustration may be caused by excessive:

    A. Pinion-bearing preload

    B. Side gear end-play

    C. Side-bearing preload

    D. Differential case/carrier runout

18. Excessive noise coming from the transfer case may be caused by all of the following EXCEPT

    A. Low transfer-case fluid level
    B. A worn transfer-case universal joint
    C. Misalignment of the transfer-case drive chain
    D. A damaged transfer-case output shaft bearing

19. Technician A says the clutch pedal freeplay adjustment sets the distance between the release bearing and the pressure-plate fingers. Technician B says a worn release bearing makes the most noise when the clutch is engaged. Who is correct?

    A. A only
    B. B only
    C. Both A and B
    D. Neither A nor B

20. Bearing preload in a transmission is:

    A. To test a bearing before it is installed
    B. The amount of pressure a bearing can handle while under a load
    C. The amount of pressure applied to a bearing upon assembly of the transmission
    D. Not adjustable

21. A four-speed manual transaxle jumps out of third gear. Technician A says the shift-rail detent spring tension on the 3-4 shift rail may be weak. Technician B says there may be excessive wear on the fourth-speed gear clutching teeth. Who is correct?

    A. A only
    B. B only
    C. Both A and B
    D. Neither A nor B

22. Technician A says the front wheel bearings on a vehicle with four-wheel drive and tapered front wheel bearings should be adjusted with 0.001–0.005 of end-play. Technician B says some late-model vehicles use unitized wheel bearings that are not adjustable. Who is correct?

    A. A only
    B. B only
    C. Both A and B
    D. Neither A nor B

23. An axle shaft is removed from the differential. Next it should be:

    A. Placed standing straight up on the floor
    B. Checked for axle flange runout after removal
    C. Left under the vehicle
    D. Placed flat in a horizontal position on the floor

24. Technician A says the transfer case can be shifted into 4L at any speed in an electronically shifted transfer case. Technician B says the transfer case can be shifted into 4L with the transmission in any gear. Who is correct?

    A. A only
    B. B only
    C. Both A and B
    D. Neither A nor B

25. Technician A says that a worn ring gear on the flywheel can cause clutch chatter. Technician B says that a worn ring gear on the flywheel can affect the starter engagement. Who is correct?

    A. A only
    B. B only
    C. Both A and B
    D. Neither A nor B

26. Which of the following components is LEAST LIKELY to be replaced when replacing the extension housing?

    A. The countergear shaft
    B. The tail-housing gasket
    C. The speedometer O-ring
    D. The output shaft seal

27. Technician A says that if the intermediate shaft gear is worn, then the mating gear on the input shaft should be replaced with the new intermediate shaft. Technician B says that if the intermediate shaft gear is worn, then the case and all of the related bearings should be checked. Who is correct?

    A. A only
    B. B only
    C. Both A and B
    D. Neither A nor B

28. Technician A says the drive-shaft carrier bearing can cause noises in neutral if the vehicle is stopped and clutch is disengaged. Technician B says that some center-support bearings need to be lubricated. Who is correct?

    A. A only
    B. B only
    C. Both A and B
    D. Neither A nor B

29. Technician A says that preload on the pinion gear should be measured with a dial indicator. Technician B says that pinion gear preload can also be measured with a flat feeler gauge. Who is correct?

    A. A only
    B. B only
    C. Both A and B
    D. Neither A nor B

30. All of the following are part of a tire inspection on a four-wheel drive vehicle EXCEPT:

    A. Aspect ratio
    B. Nominal width
    C. Load rating
    D. Brand of tire

31. A vehicle has a squealing noise coming from the clutch area when the clutch is disengaged. Technician A says the release bearing is probably dry and needs to be lubricated. Technician B says the damage has been done and the release bearing should be replaced. Who is correct?

    A. A only
    B. B only
    C. Both A and B
    D. Neither A nor B

32. A transmission is being disassembled for overhaul. Technician A says care should be taken to thoroughly inspect each component for damage as it is disassembled. Technician B says some transmission components have index marks that must be aligned when reassembled. Who is correct?

    A. A only
    B. B only
    C. Both A and B
    D. Neither A nor B

33. Technician A says that a bent shift rail should be replaced with a new shift rail. Technician B says that the shift rail may be heated and bent back into its original position using a dial indicator. Who is correct?

    A. A only
    B. B only
    C. Both A and B
    D. Neither A nor B

34. All of the following are true about removing a differential assembly EXCEPT:

    A. The axle shafts must be removed.
    B. The bearing caps should be marked to the housing.
    C. The bearing races and shim packs should not be mixed up.
    D. The pinion gear always stays in the axle housing.

35. A remote vent on a transfer case is used to do what?

    A. Add lubricating oil to the transfer case
    B. Prevent water from entering the transfer case
    C. Keep lubricant in the transfer case
    D. Increase pressure in the transfer case

36. When removing the transaxle differential from the vehicle, which of the following would be the LEAST LIKELY component to be removed?

    A. Axle shaft
    B. Transaxle
    C. Lower control arms
    D. Clutch assembly

37. When replacing a differential, Technician A says that the differential side bearings showing no damage can be reused with the new differential. Technician B says that if the differential side bearings are replaced, then the bearing races must be replaced. Who is correct?

    A. A only
    B. B only
    C. Both A and B
    D. Neither A nor B

38. On a vehicle engaged in 4WD, all of the following conditions may cause a vibration that is more noticeable when changing throttle position EXCEPT:

    A. Worn universal joints
    B. Incorrect drive shaft angles
    C. Tight drive shaft slip-joint splines
    D. Worn front-drive axle joints

39. While inspecting a reverse idler gear from a manual transmission, Technician A says that the center bore for the reverse idler should be checked for a smooth mar-free surface. Technician B says that the reverse idler gear is splined in the center bore and that the splines should be checked for excessive wear or damage. Who is correct?

    A. A only
    B. B only
    C. Both A and B
    D. Neither A nor B

40. Technician A says low range in many transfer cases is available only while in four-wheel drive mode. Technician B says many transfer cases send the power flow through a planetary gear set when in low range. Who is correct?

    A. A only
    B. B only
    C. Both A and B
    D. Neither A nor B

PREPARATION EXAM 5

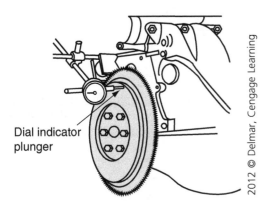

Dial indicator plunger

2012 © Delmar, Cengage Learning

1. With the dial indicator positioned as shown in the illustration, the measurement being performed is:

   A. Crankshaft runout

   B. Crankshaft end-play

   C. Crankshaft warpage

   D. Crankshaft preload

2. A manual transmission jumps out of fourth gear. Technician A says there may be excessive end-play between the fourth-speed gear and its matching synchronizer. Technician B says the detent springs on the fourth-gear shift rail may be weak. Who is correct?

   A. A only

   B. B only

   C. Both A and B

   D. Neither A nor B

3. Technician A says that the speedometer drive gear does not have to be replaced when a new speedometer cable is replaced. Technician B says that the drive gears come in different teeth numbers and using the wrong one can cause the speedometer readings to change. Who is correct?

   A. A only

   B. B only

   C. Both A and B

   D. Neither A nor B

4. A new universal joint has been installed in a vehicle. Technician A says the universal joint should be greased until grease is purged from all four caps. Technician B says that the universal joint is pre-lubricated and that drive shaft phasing can be affected by the amount of grease installed. Who is correct?

   A. A only

   B. B only

   C. Both A and B

   D. Neither A nor B

5. When adjusting the pinion depth, Technician A says to replace the collapsible spacer. Technician B says that pinion depth can also be adjusted by installing a selective pinion bearing race. Who is correct?

    A. A only
    B. B only
    C. Both A and B
    D. Neither A nor B

6. A four-wheel drive vehicle has had its tires replaced because of premature tire wear for the second time. Technician A says the tire load capacity could be too small for the vehicle. Technician B says the tires may be inflated to the incorrect pressure. Who is correct?

    A. A only
    B. B only
    C. Both A and B
    D. Neither A nor B

7. A vehicle with a manual transmission has been towed in to the shop with a no-start complaint. The technician verifies the complaint and finds that the vehicle does not crank. Technician A says to check the clutch safety switch for an open circuit. Technician B says the clutch switch could be out of adjustment. Who is correct?

    A. A only
    B. B only
    C. Both A and B
    D. Neither A nor B

8. When diagnosing a DTC for the output-shaft speed sensor, which of the following should NOT be done next after reading the code?

    A. Check the service manual.
    B. Install a new sensor.
    C. Inspect the connector.
    D. Inspect the harness for continuity.

9. A transaxle has transaxle fluid leaking from the output shaft seal on an all-wheel drive vehicle; the technician removes the seal for inspection and cannot see any problem with the seal. Technician A says to check for a damaged yoke seal surface. Technician B says the vent could be plugged. Who is correct?

    A. A only
    B. B only
    C. Both A and B
    D. Neither A nor B

10. A rear-wheel drive vehicle is being diagnosed for a speed-related vibration. Technician A says most joint working angles should not exceed 3 degrees. Technician B says two joint working angles on a common shaft should cancel each other out within 1 degree. Who is correct?

    A. A only
    B. B only
    C. Both A and B
    D. Neither A nor B

11. Differential side-bearing preload is being adjusted. Technician A says that some RWD differentials require the selection of the correct shim thickness to adjust the side-bearing preload. Technician B says that some RWD differentials use threaded adjusters to set side-bearing preload. Who is correct?

    A. A only
    B. B only
    C. Both A and B
    D. Neither A nor B

12. A transfer case is being removed for servicing. Technician A says the transfer case and transmission must be removed as a unit. Technician B says all yokes and flanges should be marked for correct reassembly. Who is correct?

    A. A only
    B. B only
    C. Both A and B
    D. Neither A nor B

13. When replacing a clutch, the pressure plate is found to have small cracks. Technician A says the pressure plate should be replaced. Technician B says the pressure plate should be resurfaced and then installed. Who is correct?

    A. A only
    B. B only
    C. Both A and B
    D. Neither A nor B

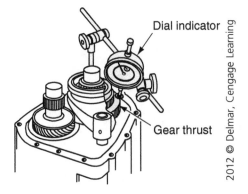

14. Technician A says transmission preload is being measured in the illustration. Technician B says if the measurement is excessive, then the thrust washer can be replaced to correct the measurement.

    A. A only
    B. B only
    C. Both A and B
    D. Neither A nor B

15. A four-speed transaxle has a clunking noise when driven in first gear and reverse gear. Technician A says that the gear on the countershaft could be the problem. Technician B says that the reverse idler gear could be the problem. Who is correct?

    A. A only

    B. B only

    C. Both A and B

    D. Neither A nor B

16. A clicking noise is heard on a front-wheel drive vehicle while turning a corner. The cause of this problem could be a bad:

    A. Inner constant-velocity joint

    B. Front axle

    C. Wheel bearing

    D. Outer constant-velocity joint

17. A limited-slip differential is not working properly. Which of the following would be the LEAST LIKELY cause?

    A. Worn friction disc

    B. Weak preload spring

    C. Wrong oil additive used

    D. Incorrect shim used on the side bearing

18. On a four-wheel drive vehicle with a straight front axle, front universal joint working angle can be corrected by:

    A. Shimming the front axle housing

    B. Installing a shorter drive shaft

    C. The joint working angles cannot be changed

    D. Installing a 4-inch body lift

19. A defective or worn pilot bearing may cause a rattling and growling noise under which condition?

    A. The engine is idling with the clutch engaged.

    B. The engine is idling and the clutch pedal is disengaged.

    C. The engine is idling and the clutch is fully disengaged.

    D. The engine is idling and the clutch pedal is fully disengaged.

20. Technician A says that many transmissions with internal linkage have no internal linkage adjustments. Technician B says that only transmissions with external linkage can be adjusted. Who is correct?

    A. A only

    B. B only

    C. Both A and B

    D. Neither A nor B

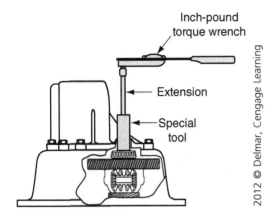

21. Technician A says that differential-bearing preload is being measured in the above illustration. Technician B says the final side-gear fastener torque is being applied. Who is correct?

    A. A only
    B. B only
    C. Both A and B
    D. Neither A nor B

22. When installing a replacement U-joint that has a grease zerk in the cross, the grease zerk should point toward the:

    A. Yoke
    B. Drive shaft
    C. Engine
    D. Differential

23. Technician A says to rotate the axle slowly when checking the rear-axle runout. Technician B says to use a dial indicator to measure runout. Who is correct?

    A. A only
    B. B only
    C. Both A and B
    D. Neither A nor B

24. A vehicle with full-time, four-wheel drive has a different size tire on every wheel. Technician A says this could cause all three differentials to fail prematurely. Technician B says if the owner puts in special additive, then the differentials will be OK. Who is correct?

    A. A only
    B. B only
    C. Both A and B
    D. Neither A nor B

25. Technician A says that clutch chatter could be caused by an uneven flywheel. Technician B says that oil on the clutch disc can cause clutch chatter. Who is correct?

    A. A only
    B. B only
    C. Both A and B
    D. Neither A nor B

26. A transmission will not engage in first gear when shifted. Which of these is the LEAST LIKELY cause?

    A. A broken synchronizer sleeve

    B. Misadjusted linkage

    C. A broken shift fork

    D. A broken tooth on the first gear

27. Technician A says that the input shaft end-play does not need to be checked before disassembling a transaxle. Technician B says to always rotate the input shaft to check turning effort before disassembling the transaxle. Who is correct?

    A. A only

    B. B only

    C. Both A and B

    D. Neither A nor B

28. Technician A says that drive shaft runout should be checked with a digital caliper at the middle of the drive shaft. Technician B says that a dial indicator should be set at the differential end of the drive shaft to check runout. Who is correct?

    A. A only

    B. B only

    C. Both A and B

    D. Neither A nor B

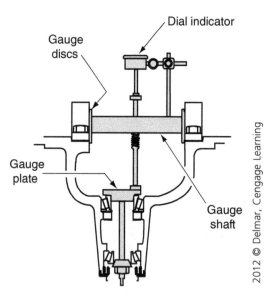

29. In the figure shown, after the dial indicator is rotated to the zero position with the stem on the gauge plate and then moved off the gauge plate, the dial indicator pointer moves 0.057 inches (1.45 mm) counterclockwise and the pinion gear is marked −4. The proper pinion depth shim is:

    A. 0.043 inch (1.09 mm)

    B. 0.042 inch (1.07 mm)

    C. 0.039 inch (0.99 mm)

    D. 0.041 inch (1.04 mm)

30. When removing a transfer case from the vehicle, which of the following components listed is LEAST LIKELY to be disconnected or removed?

    A. The front drive shaft
    B. The rear drive shaft
    C. The transmission
    D. The linkage

31. Technician A says constant-running release bearings are used with hydraulically controlled clutches. Technician B says on a push-type pressure plate, the release bearing moves away from the pressure plate to disengage the clutch. Who is correct?

    A. A only
    B. B only
    C. Both A and B
    D. Neither A nor B

32. The clearance on the third-speed gear blocking ring is less than specified. Technician A says this may result in noise while driving in third gear. Technician B says this problem may cause hard shifting into third gear. Who is correct?

    A. A only
    B. B only
    C. Both A and B
    D. Neither A nor B

33. Technician A says that most transaxle cases are sealed with room temperature vulcanizing (RTV) sealant. Technician B says that some transaxle cases have paper gaskets. Who is correct?

    A. A only
    B. B only
    C. Both A and B
    D. Neither A nor B

34. To measure differential case/carrier runout, which of the following tools would be used?

    A. Micrometer
    B. Feeler gauge
    C. Dial indicator
    D. Torque wrench

35. A vehicle with four-wheel drive is in the shop for front-axle seal leakage. Upon draining the fluid, the technician notices that the fluid is milky and watery. Technician A says the condition could be from normal operation in high water. Technician B says the vent location should be checked to make sure it has not broken and the hose is not split. Who is correct?

    A. A only
    B. B only
    C. Both A and B
    D. Neither A nor B

36. Technician A says that a transmission that is binding internally should be installed and allowed to break-in. Technician B says that if the transmission gets hung up during installation, then it should be drawn up the rest of the way using the bolts. Who is correct?

    A. A only
    B. B only
    C. Both A and B
    D. Neither A nor B

37. Each of the following statements about manual transaxle shift linkages are true EXCEPT:

    A. Shift linkages may be cable operated.
    B. Shift linkage cables are longer on transaxles than on rear-wheel drive transmissions.
    C. Shift linkage may be rod-operated.
    D. All shift linkages are adjustable.

38. Technician A says the collapsible pinion-shaft spacer may be reused if the differential is disassembled and overhauled only once. Technician B says that after proper pinion bearing preload is set, pinion depth is adjusted by backing off the pinion flange nut one quarter turn. Who is correct?

    A. A only
    B. B only
    C. Both A and B
    D. Neither A nor B

39. Technician A says low oil level in the transfer case is a sign of a leak. Technician B says overfilling a transfer case can cause an increase in fluid temperature. Who is correct?

    A. A only
    B. B only
    C. Both A and B
    D. Neither A nor B

40. Technician A says that all linkages on an electronic-shift transfer case are external. Technician B says that an open in the shift switch circuit will prevent the transfer case from engaging. Who is correct?

    A. A only
    B. B only
    C. Both A and B
    D. Neither A nor B

PREPARATION EXAM 6

1. Which of the following is the LEAST Likely cause of clutch chatter?

    A. Weak clutch-disc torsional springs

    B. A rear main seal leaking

    C. A transmission input shaft seal leaking

    D. Excessive input shaft end-play

2. Uneven wear of the extension-housing bushing will most likely be caused by:

    A. Worn speedometer drive and drive gears

    B. Excessive transmission main shaft end-play

    C. A plugged transmission vent opening

    D. Uneven mating surfaces on the extension housing

3. Technician A says an improper shift linkage adjustment may cause the transaxle to jump out of gear under a load. Technician B says a broken engine mount may cause the transaxle to jump out of gear. Who is correct?

    A. A only

    B. B only

    C. Both A and B

    D. Neither A nor B

4. To measure drive shaft runout, the technician should do all of the following EXCEPT

    A. Place the vehicle in neutral.

    B. Use a micrometer to check the drive shaft for parallelism.

    C. Clean the drive shaft at the spot where the measurement will be taken.

    D. Jack up the vehicle and place on jack stands.

5. Which of the following would be the cause for the brakes to be contaminated by rear-end oil?

    A. A tight bearing-retainer ring

    B. The worn oil-seal collar

    C. The axle shaft bearings are worn

    D. The axle bearing retainer plate distorted

6. The most likely cause of a manual-shift transfer case not shifting into four-wheel drive (4WD) is which of the following?

    A. The transfer case oil level is low.

    B. The rear drive shaft universal joints are bound up.

    C. The shift linkage is out of adjustment.

    D. The electronic shift motor is bad.

7. A bearing-type noise begins to come from the clutch and transmission area of a vehicle just as the clutch is almost completely disengaged. There is no noise when the clutch pedal is initially depressed. Technician A says that the clutch pilot bearing may be worn out. Technician B says that the release bearing may be worn out. Who is correct?

    A. A only

    B. B only

    C. Both A and B

    D. Neither A nor B

8. A manual transmission, when in neutral, has a growling noise with the engine idling and the clutch engaged. The noise disappears when the clutch pedal is depressed. This noise could be caused by what?

    A. Pilot bearing in the crankshaft

    B. Input shaft and pilot bearing contact area

    C. Input shaft bearing

    D. Mainshaft bearing

9. The needle bearings between the output shaft and the output shaft gears are scored and blue from overheating on a transaxle. Technician A says the transaxle may have been filled with the wrong lubricant at its last service. Technician B says the oil passage that supplies oil to those bearings may be restricted or plugged. Who is correct?

    A. A only

    B. B only

    C. Both A and B

    D. Neither A nor N

10. Technician A says that a clunking noise while cornering could be caused by a bad constant-velocity (CV) joint. Technician B says a CV joint could also make noise when the vehicle is traveling straight. Who is correct?

    A. A only

    B. B only

    C. Both A and B

    D. Neither A nor B

11. Technician A says that if the differential is overfilled, then it can cause axle seals to leak. Technician B says that worn axle bearings can cause the axle seals to leak. Who is correct?

    A. A only

    B. B only

    C. Both A and B

    D. Neither A nor B

12. Mechanical transfer case linkage is being adjusted. Technician A says the transfer case lever should be placed in four-wheel drive high. Technician B says the transfer case levers are indexed using a pin or drill bit. Who is correct?

    A. A only
    B. B only
    C. Both A and B
    D. Neither A nor B

13. When checking flywheel runout, which one of the following is LEAST LIKELY to affect runout?

    A. Crankshaft end-play
    B. Ring gear wear
    C. Flywheel fastener torque
    D. Dial indicator stylus placement

14. Technician A says that when installing a cork gasket, no added sealant is required. Technician B says that rubber gaskets should be installed without any added sealant. Who is correct?

    A. A only
    B. B only
    C. Both A and B
    D. Neither A nor B

15. When draining the fluid from a manual transaxle, a gold-colored material is seen in the fluid. The most likely cause is a worn

    A. Blocker ring
    B. Speed gear
    C. Reverse Idler
    D. Countershaft

16. Technician A says that a strobe light and screw-type hose clamps can be used to balance a drive shaft in the vehicle. Technician B says some drive shafts have match marks for balance. Who is correct?

    A. A only
    B. B only
    C. Both A and B
    D. Neither A nor B

17. When draining a limited-slip differential, a paper-like material is found in the fluid. What is the most likely cause?

    A. A worn limited-slip synchronizer
    B. Worn limited-slip disc
    C. Damaged axle seal
    D. Damaged pinion seal

18. A four-wheel drive vehicle with front half shafts has a torn CV boot. What is the proper way to repair the boot?

    A. Replace the axle assembly.
    B. Replace the CV joint and boot.
    C. Repack the CV joint, and install a quick boot (split boot).
    D. Repack the CV joint, and reseal the boot with an approved sealer.

19. Technician A says that the freeplay at the clutch pedal should be measured before attempting to adjust the clutch. Technician B says if the clutch is slipping badly, then it can be corrected by adjusting the clutch. Who is correct?

    A. A only

    B. B only

    C. Both A and B

    D. Neither A nor B

20. Transmission end-play can be adjusted by all of the following EXCEPT

    A. Different length bolts

    B. Thrust washers

    C. Thrust shims

    D. Selective snap rings

21. All of the following about manual transaxle shift linkage is true EXCEPT

    A. Many transmissions use steering-column-mounted gear shift levers.

    B. Shift linkage can be cable operated.

    C. Shift linkage cables are longer for a transaxle than those on a rear-wheel drive transmission.

    D. Shift linkage can be internal.

22. Which statement is true concerning a center-support bearing?

    A. It is usually a sealed bearing and is maintenance-free.

    B. It should be greased like a universal joint.

    C. It is part of the drive shaft and cannot be replaced separately.

    D. It is used on all rear-wheel drive vehicles.

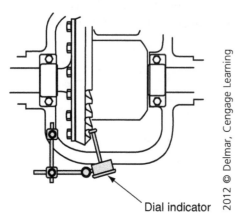

Dial indicator

2012 © Delmar, Cengage Learning

23. In the figure shown, a technician is measuring

    A. Ring-gear runout

    B. Ring-gear backlash

    C. Bearing preload

    D. Pinion gear backlash

24. Front universal joints are being replaced. Technician A says the new U-joints will still need to be greased after assembly. Technician B says if the universal joint is installed incorrectly, then the grease fitting cannot be reached with a grease gun. Who is correct?

    A. A only
    B. B only
    C. Both A and B
    D. Neither A nor B

25. Technician A says sagged transmission mounts may cause improper drive shaft angles on a rear-wheel drive car. Technician B says improper drive shaft angles may cause a constant-speed vibration when the vehicle is accelerated and decelerated. Who is correct?

    A. A only
    B. B only
    C. Both A and B
    D. Neither A nor B

26. Excessive input shaft end-play in a 4-speed transmission may cause the transmission to jump out of

    A. First gear
    B. Second gear
    C. Fourth gear
    D. Reverse gear

27. A transaxle shifts normally into all forward gears, but it will not shift into reverse gear; there is no evidence of noise while attempting this shift. Technician A says the reverse shifter fork may be broken. Technician B says the reverse idler gear teeth may be worn. Who is correct?

    A. A only
    B. B only
    C. Both A and B
    D. Neither A nor B

28. When replacing an axle boot, which component is LEAST LIKELY to be visually inspected?

    A. The constant-velocity joint
    B. A wheel bearing
    C. An axle shaft
    D. Axle seals

29. When replacing the ring and pinion gears, all of the following must be replaced EXCEPT

    A. Spider gears
    B. Pinion seal
    C. Collapsible spacer
    D. Axle seals

30. A four-wheel drive vehicle has a jerking feeling in the steering wheel when turning sharply with the four-wheel drive engaged. Technician A says this could be normal for many four-wheel drive vehicles. Technician B says if the vehicle has CV joints in the front axle, then this would not be felt as much. Who is correct?

    A. A only

    B. B only

    C. Both A and B

    D. Neither A nor B

31. Technician A says that a clutch may slip when it is out of adjustment. Technician B says that a transmission may grind when the clutch is out of adjustment. Who is correct?

    A. A only

    B. B only

    C. Both A and B

    D. Neither A nor B

32. Technician A says that a transmission that is hard to shift in gear may have a problem with the linkage not being lubricated. Technician B says that this problem may be caused by too strong a pressure plate installed in the vehicle's clutch system. Who is correct?

    A. A only

    B. B only

    C. Both A and B

    D. Neither A nor B

33. Technician A says that if side gears are damaged, then noise would only be heard when the vehicle is turned. Technician B says that side gears are shimmed for proper clearance. Who is correct?

    A. A only

    B. B only

    C. Both A and B

    D. Neither A nor B

34. Technician A says if the ring-gear tooth contact pattern is low at the toe of the tooth, then the pinion gear should be moved toward the ring gear. Technician B says if the pinion gear teeth have low flank contact on the ring-gear teeth, then the pinion gear should be moved toward the ring gear. Who is correct?

    A. A only

    B. B only

    C. Both A and B

    D. Neither A nor B

35. A vehicle with four-wheel drive is in the shop for front-axle seal leakage. Upon draining the fluid, the technician notices that the fluid is milky and watery. Technician A says the condition could be from normal operation in high water. Technician B says the vent location should be checked to make sure it has not broken and that the hose is not split. Who is correct?

    A. A only

    B. B only

    C. Both A and B

    D. Neither A nor B

36. A car with a 4-speed transmission has a growling noise coming from the transmission in all gears except fourth. In fourth, the noise is almost completely gone. Technician A says the input shaft bearing could cause the noise. Technician B says the countershaft bearing could be causing the noise. Who is correct?

    A. A only
    B. B only
    C. Both A and B
    D. Neither A nor B

37. A transaxle has transaxle fluid leaking from the output shaft seal on an all-wheel drive vehicle. The technician removes the seal for inspection and cannot see any problem with the seal. Technician A says to check for a damaged yoke seal surface. Technician B says the vent could be plugged. Who is correct?

    A. A only
    B. B only
    C. Both A and B
    D. Neither A nor B

38. A vehicle with rear-wheel drive needs its spider gears replaced. Technician A says that some RWD axles require the removal of the pinion (spider) gear shaft in order to replace the spider gears. Technician B says that some RWD axles are held in place by bolts at the outer axle flange. Who is correct?

    A. A only
    B. B only
    C. Both A and B
    D. Neither A nor B

39. Technician A says an unbalanced drive shaft causes a vibration that increases as vehicle speed increases. Technician B says an unbalanced drive shaft causes a vibration that decreases as vehicle speed increases. Who is correct?

    A. A only
    B. B only
    C. Both A and B
    D. Neither A nor B

40. A vehicle with all-wheel drive has a failed center differential. Technician A says mismatched tires could have caused this. Technician B says tires of the same size, but smaller than required, could have caused this. Who is correct?

    A. A only
    B. B only
    C. Both A and B
    D. Neither A nor B

INTRODUCTION

Included in this section are the answer keys for each preparation exam, followed by individual, detailed answer explanations and a reference identifying the designated task area being assessed by each specific question. This additional reference information may prove useful if you need to refer back to the task list located in Section 4 of this book for additional support.

PREPARATION EXAM 1—ANSWER KEY

1.	C	21.	C
2.	C	22.	A
3.	D	23.	D
4.	C	24.	A
5.	B	25.	B
6.	C	26.	C
7.	C	27.	C
8.	C	28.	C
9.	A	29.	D
10.	C	30.	D
11.	A	31.	C
12.	B	32.	A
13.	A	33.	C
14.	C	34.	C
15.	A	35.	D
16.	A	36.	A
17.	A	37.	C
18.	B	38.	B
19.	C	39.	C
20.	C	40.	C

PREPARATION EXAM 1—EXPLANATIONS

1. A manual transmission grinds when attempting to select first or reverse gear only. Which of these could be the cause?

    A. A dry release bearing

    B. A chipped speed gear

    C. A warped clutch disc

    D. Insufficient clutch free travel

    TASK A.1

    Answer A is incorrect. A dry release bearing would make noise, not gear selection problems.

    Answer B is incorrect. A chipped speed gear would possibly make a noise, not a gear selection problem.

    Answer C is correct. A warped clutch disc could cause a clutch spin-up problem that could cause the transmission to grind when selecting first or reverse gears. The transmission would not grind when selecting the remaining gears because the vehicle is in motion when they are selected.

    Answer D is incorrect. Insufficient clutch free travel would cause clutch slippage.

2. Technician A says broken transmission mount can cause excessive driveline angles. Technician B says an oil-soaked transmission mount should be inspected very closely for rubber deterioration and swelling. Who is correct?

    A. A only

    B. B only

    C. Both A and B

    D. Neither A nor B

    TASK A.10

    Answer A is incorrect. Technician B is also correct.

    Answer B is incorrect. Technician A is also correct.

    Answer C is correct. Both technicians are correct. A broken transmission mount can cause variations in driveline angle that could cause vibrations. Petroleum products like motor oil and transmission fluid can cause premature failure of rubber components, such as power train mounts.

    Answer D is incorrect. Both technicians are correct.

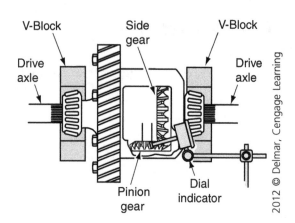

TASK E.2.3

3. Which of these is the technician measuring in the illustration?

    A.   Ring-gear backlash

    B.   Differential end-play

    C.   Side gear backlash

    D.   Pinion gear backlash

    Answer A is incorrect. Ring-gear backlash would be measured on the ring gear.

    Answer B is incorrect. Differential end-play is not being measured. Differential end-play is measured with the differential installed in the transaxle with the aid of some special tools.

    Answer C is incorrect. Side gear backlash would be measured on the side gear.

    Answer D is correct. The dial indicator stylus is on the pinion gear to display pinion gear backlash.

TASK D.6

4. A rear-wheel drive vehicle is being diagnosed for a speed-related vibration. Technician A says most joint working angles should not exceed 3 degrees. Technician B says two joint working angles on a common shaft should cancel each other out within 1 degree. Who is correct?

    A.   A only

    B.   B only

    C.   Both A and B

    D.   Neither A nor B

    Answer A is incorrect. Technician A is also correct.

    Answer B is incorrect. Technician B is also correct.

    Answer C is correct. U-joint working angles should be at least ½ degree and no more than 3 degrees unless specified by the manufacturer. They should cancel each other out within 1 degree.

    Answer D is incorrect. Both technicians are correct.

5. A transfer case is being removed for servicing. Technician A says the transfer case and transmission must be removed as a unit. Technician B says all yokes and flanges should be marked for correct reassembly. Who is correct?

TASK F.3

   A. A only
   B. B only
   C. Both A and B
   D. Neither A nor B

   Answer A is incorrect. Transfer cases and transmissions do not have to be removed as a unit.

   Answer B is correct. Only technician B is correct. All yokes and flanges should be marked for correct reassembly.

   Answer C is incorrect. Only technician B is correct.

   Answer D is incorrect. Technician B is correct.

6. A vehicle with limited slip differential is being diagnosed for chattering on a turn. Technician A says the differential may have the incorrect type of lubricant in it. Technician B says the last technician that serviced the differential may have not added the special limited-slip additive. Who is correct?

TASK C.16

   A. A only
   B. B only
   C. Both A and B
   D. Neither A nor B

   Answer A is incorrect. Technician B is also correct.

   Answer B is incorrect. Technician A is also correct.

   Answer C is correct. Both technicians are correct. A limited differential that chatters on turns could be as a result of incorrect fluid being used or due to the lack of a limited-slip differential additive.

   Answer D is incorrect. Both technicians are correct.

7. A vehicle with a manual transmission has been towed into the shop with a no-start complaint. The technician verifies the complaint and finds that the vehicle will not crank. Technician A says to check the clutch safety switch for an open circuit. Technician B says the clutch switch could be out of adjustment. Who is correct?

TASK A.2

   A. A only
   B. B only
   C. Both A and B
   D. Neither A nor B

   Answer A is incorrect. Technician B is also correct.

   Answer B is incorrect. Technician B is also correct.

   Answer C is correct. Both technicians are correct. The clutch safety switch prevents starter engagement unless the clutch is disengaged. A defective or out-of-adjustment clutch switch could cause a starter engagement problem.

   Answer D is incorrect. Both technicians are correct.

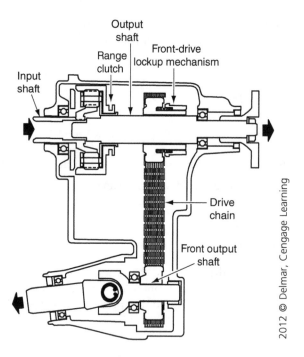

Input shaft

Range clutch

Output shaft

Front-drive lockup mechanism

Drive chain

Front output shaft

2012 © Delmar, Cengage Learning

TASK F.1

8. Technician A says if the drive chain shown in the illustration were to break, then the vehicle would still have power flow to the rear wheels. Technician B says low range can be achieved with a broken drive chain. Who is correct?

    A. A only

    B. B only

    C. Both A and B

    D. Neither A nor B

    Answer A is incorrect. Technician B is also correct.

    Answer B is incorrect. Technician A is also correct.

    Answer C is correct. Both technicians are correct. The drive chain in this transfer case only transfers power to the front axle; rear-wheel drive function would not be affected. High/low range is achieved by engaging and disengaging the range clutch. Still it would be possible to shift the vehicle to low range.

    Answer D is incorrect. Both technicians are correct.

TASK B.10

9. During a transmission overhaul, a technician finds a chipped tooth on the countershaft. Technician A says the countershaft and mating speed gear should be replaced. Technician B says only the damaged countershaft should be replaced. Who is correct?

    A. A only

    B. B only

    C. Both A and B

    D. Neither A nor B

    Answer A is correct. Only technician A is correct. Any time a gear is replaced in a transmission, its mating gear should also be replaced.

    Answer B is incorrect. Any time a gear is replaced in a transmission, its mating gear should also be replaced.

    Answer C is incorrect. Only technician A is correct.

    Answer D is incorrect. Technician A is correct.

10. A vehicle with four-wheel drive and tapered front-wheel bearing is being serviced. Technician A says the front wheel bearings should adjusted with 0.001–0.005 inches of end-play. Technician B says some late-model vehicles use unitized wheel bearings. Who is correct?

TASK D.7

   A. A only

   B. B only

   C. Both A and B

   D. Neither A nor B

   Answer A is incorrect. Technician B is also correct.

   Answer B is incorrect. Technician A is also correct.

   Answer C is correct. Both technicians are correct. Tapered front wheel bearings should be adjusted with a slight end-play of 0.001–0.005 inches. Some late-model vehicles, however, use unitized wheel bearing.

   Answer D is incorrect. Both technicians are correct.

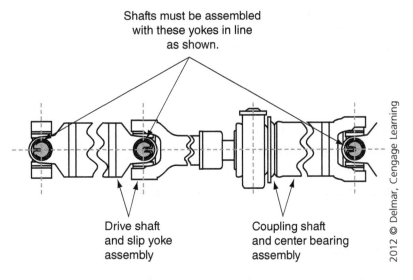

Shafts must be assembled
with these yokes in line
as shown.

Drive shaft
and slip yoke
assembly

Coupling shaft
and center bearing
assembly

2012 © Delmar, Cengage Learning

11. A drive shaft like the one shown in the illustration is reassembled after U-joint replacement without lining up the yokes. Technician A says this could cause driveline vibration. Technician B says the yokes must be in line for clearance. Who is correct?

TASK D.1

   A. A only

   B. B only

   C. Both A and B

   D. Neither A nor B

   Answer A is correct. Only technician A is correct. To prevent vibration, the yoke must be in line so that the U-joint angles can cancel each other out.

   Answer B is incorrect. Lining the yokes up is to aid vibration control, not to provide clearance.

   Answer C is incorrect. Only technician A is correct.

   Answer D is incorrect. Technician A is correct.

TASK A.4

12. A vehicle has a squealing noise coming from the clutch area when the clutch is disengaged. Technician A says the release bearing is probably dry and needs to be lubricated. Technician B says the damage has been done and the release bearing should be replaced. Who is correct?

A. A only

B. B only

C. Both A and B

D. Neither A nor B

Answer A is incorrect. Most release bearings cannot be greased; once the bearing starts making noise, the damage has been done.

Answer B is correct. Only technician B is correct. Once the bearing starts making noise, the damage has been done.

Answer C is incorrect. Only technician B is correct

Answer D is incorrect. Technician B is correct.

TASK A.6

13. A vehicle needs to have its pilot bearing replaced (bronze bushing type). Technician A advises to make sure you check pilot-bearing bore runout. Technician B says it is necessary to lubricate the pilot bearing with high-quality white lithium grease. Who is correct?

A. A only

B. B only

C. Both A and B

D. Neither A nor B

Answer A is correct. Only technician A is correct. Any time the pilot bearing is replaced, it is a good idea to check the pilot bearing bore runout.

Answer B is incorrect. Most manufacturers only require a small amount of high-temperature grease.

Answer C is incorrect. Only technician A is correct.

Answer D is incorrect. Technician A is correct.

TASK E.2.3

14. Differential side-bearing preload is being adjusted. Technician A says that some RWD differentials require the selection of the correct shim thickness to adjust the side-bearing preload. Technician B says that some RWD differentials use threaded adjusters to set side-bearing preload. Who is correct?

A. A only

B. B only

C. Both A and B

D. Neither A nor B

Answer A is incorrect. Technician B is also correct.

Answer B is incorrect. Technician A is also correct.

Answer C is correct. Both technicians are correct. On some differentials, side-bearing preload is determined by shim thickness behind the side bearing. Some differentials use threaded adjusters to set side-bearing preload.

Answer D is incorrect. Both technicians are correct.

15. A transaxle has had repeated drive axle seal leakage and replacement. Technician A says this problem may be caused by a plugged transaxle vent. Technician B says this problem may be caused by a worn outer-drive axle joint. Who is correct?

TASK C.3

    A. A only

    B. B only

    C. Both A and B

    D. Neither A nor B

Answer A is correct. Only technician A is correct. A plugged vent can cause enough internal pressure to force fluid past a seal.

Answer B is incorrect. An outer drive axle joint does not come into contact with the drive seal.

Answer C is incorrect. Only technician A is correct.

Answer D is incorrect. Technician A is correct.

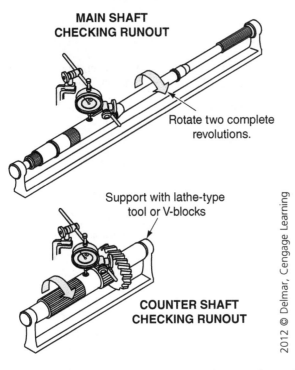

MAIN SHAFT CHECKING RUNOUT

Rotate two complete revolutions.

Support with lathe-type tool or V-blocks

COUNTER SHAFT CHECKING RUNOUT

2012 © Delmar, Cengage Learning

16. Refer to the supplied illustration. Technician A says if the mainshaft runout is above specifications, then the mainshaft must be replaced. Technician B says the mainshaft can be straightened using an acetylene torch and hydraulic press. Who is correct?

TASK B.8

    A. A only

    B. B only

    C. Both A and B

    D. Neither A nor B

Answer A is correct. Only technician A is correct. A mainshaft with excessive runout must be replaced.

Answer B is incorrect. A torch and press should not be used to repair the mainshaft. The excessive heat will change the temper of the metal in the mainshaft.

Answer C is incorrect. Only technician A is correct.

Answer D is incorrect. Technician A is correct.

TASK E.3.3

17. A limited-slip differential with clutch packs is being overhauled. Technician A says the clutch pack discs should be installed alternately (steel, fiber, steel). Technician B says the clutch pack discs should be installed with the steel discs on the right and fiber discs on the left. Who is correct?

   A.   A only
   B.   B only
   C.   Both A and B
   D.   Neither A nor B

   Answer A is correct. Only technician A is correct. A limited-slip differential that uses a clutch pack should have the fiber discs soaked in differential lube, then installed alternately (steel, fiber, steel).

   Answer B is incorrect. The clutch discs are not installed with all steel discs on the right and all fiber discs on the left. They are installed alternately (steel, fiber, steel).

   Answer C is incorrect. Only technician A is correct.

   Answer D is incorrect. Technician A is correct.

TASK A.3

18. In a hydraulic clutch system, the clutch fails to disengage properly when the clutch pedal is fully depressed. The cause of this problem might be which of the following?

   A.   Less-than-specified clutch pedal freeplay
   B.   Air in the clutch hydraulic system
   C.   Worn clutch facings
   D.   A scored pressure plate

   Answer A is incorrect. A hydraulic clutch system does not require freeplay.

   Answer B is correct. Air in the system will prevent the clutch from disengaging properly.

   Answer C is incorrect. If the clutch facings were worn, then it would have no effect on clutch disengagement.

   Answer D is incorrect. A scored pressure plate would not cause clutch disc disengagement problems.

TASK B.1

19. A truck with a 4-speed transmission has a growling noise coming from the transmission in all gears except fourth. In fourth, the noise is almost completely gone. Technician A says the input shaft bearing could cause the noise. Technician B says the countershaft bearing could be causing the noise. Who is correct?

   A.   A only
   B.   B only
   C.   Both A and B
   D.   Neither A nor B

   Answer A is incorrect. Technician B is also correct.

   Answer B is incorrect. Technician A is also correct.

   Answer C is correct. Both technicians are correct. A faulty input shaft bearing or faulty countershaft bearing would make noise in all gears except fourth (direct drive) because in direct drive the gear ratio is 1:1. Although the input shaft and countershafts are turning, the power flow is directly through the input to mainshaft, and the bearings are not loaded as much.

   Answer D is incorrect. Both technicians are correct.

20. A transaxle has been disassembled for intermediate shaft bearing replacement. Technician A says snap rings should not be reused and that new ones should be installed. Technician B says bearings that pass inspection can be reused as long as they were removed using a puller and not driven off with a hammer. Who is correct?

TASK C.7

   A.   A only
   B.   B only
   C.   Both A and B
   D.   Neither A nor B

Answer A is incorrect. Technician B is also correct.

Answer B is incorrect. Technician A is also correct.

Answer C is correct. Both technicians are correct. When overhauling or disassembling a transaxle for a bearing replacement, any snap ring removed should be replaced with a new one. Bearings can be reused if they do not show any signs of wear and were not driven off with a hammer. Using a puller does not compromise the bearing integrity, but driving them on and off does.

Answer D is incorrect. Both technicians are correct.

21. A vehicle with automatic locking hubs is being diagnosed for slow engagement. Technician A says that many vehicles with automatic locking hubs require as much as a full tire rotation in a specified direction in order for the hub to engage and disengage. Technician B says the driver does not have to leave the vehicle to engage the automatic locking hubs. Who is correct?

TASK F.11

   A.   A only
   B.   B only
   C.   Both A and B
   D.   Neither A nor B

Answer A is incorrect. Technician B is also correct.

Answer B is incorrect. Technician A is also correct.

Answer C is correct. Both technicians are correct. The main advantage of automatic locking hubs is that the driver does not need to leave the vehicle to activate four-wheel drive (4WD) or drive the vehicle in 2WD with the front axle engaged. The disadvantage with this system is that most designs require the vehicle to move some distance (usually a whole wheel turn in a specific direction) in order for the hubs to engage or disengage.

Answer D is incorrect. Both technicians are correct.

22. All of the following statements are true about removing and installing axle shafts EXCEPT:

TASK E.4.3

   A.   Inspect the shaft near the seal area and repair with fine sand paper, as needed.
   B.   The axle shaft seals should be replaced, not reused.
   C.   The axle shaft should be stood straight up in a vertical position when removed.
   D.   Some axles require a slide hammer to remove them.

Answer A is correct. If the shaft has damage in the seal area, then it should be replaced.

Answer B is incorrect. It is always a good practice to replace the axle seals while the axle shafts are removed.

Answer C is incorrect. Always stand the axle shaft on its end so that the splines do not contact the floor, which could cause damage.

Answer D is incorrect. Some require the use of a slide hammer to remove them from the differential.

TASK C.8

23. The second-speed gear clutch teeth and blocking ring teeth are badly worn on a transaxle. This problem may cause which of the following conditions?

    A.   A growling noise while driving in second gear

    B.   A vibration while accelerating in second gear

    C.   Hard shifting in second and third gear

    D.   The transaxle to jump out of second gear

    Answer A is incorrect. The clutch teeth only would cause noise when shifting into second gear.

    Answer B is incorrect. A vibration in second gear would be an indication of gear problems.

    Answer C is incorrect. Difficult shifting would happen in second gear only.

    Answer D is correct. Worn gear clutch teeth will not allow the synchronizer sleeve to grip the gear and will allow the transmission to slip out of gear. The synchronizer is used to match the gear speed with the main or intermediate shaft speed for smooth gear engagement. As clutch teeth on the speed gear and the blocker ring wear, it becomes easier for the transaxle to jump out of gear.

TASK B.13

24. An extension housing has burrs and gouges on the mating surface. Technician A says that if they are not excessive, they can be repaired with a file. Technician B says that the mating surfaces are machined surfaces and they should only be repaired with fine grit sandpaper. Who is correct?

    A.   A only

    B.   B only

    C.   Both A and B

    D.   Neither A nor B

    Answer A is correct. Only technician A is correct. The gasket or sealer will seal the mating surfaces if the burrs have been filed off.

    Answer B is incorrect. The gasket or sealer will seal the mating surfaces if the burrs have been filed off, and sand paper can be used to make surface repairs, but saying it should only be repaired with fine grit sand paper is incorrect.

    Answer C is incorrect. Only technician A is correct.

    Answer D is incorrect. Technician A is correct.

TASK F.10

25. A remote vent on a differential on a 4WD vehicle is used to:

    A.   Increase pressure in the differential

    B.   Keep moisture out of the differential

    C.   Keep lubricant from coming out of the differential

    D.   Add lubricant to the differential

    Answer A is incorrect. The remote vent tube on a differential does not increase pressure; it vents pressure.

    Answer B is correct. The remote vent tube is located above a point where moisture could enter. Because 4WD vehicles are designed for off-road use, the vent is located at a much higher level than the vent on a vehicle designed for on-road use only.

    Answer C is incorrect. The remote vent tube on a differential does not keep lubricant from coming out of the differential. It must be able to allow pressure to vent. If a differential was grossly overfilled, then it is possible to see differential lube leak from the vent.

    Answer D is incorrect. The remote vent tube on a differential is not used to add lubricant to the differential.

26. Technician A says that if the differential is overfilled, then it will cause axle seals to leak. Technician B says that worn axle bearings will cause axle seals to leak. Who is correct?

TASK E.1.1

    A. A only

    B. B only

    C. Both A and B

    D. Neither A nor B

Answer A is incorrect. Technician B is also correct.

Answer B is incorrect. Technician A is also correct.

Answer C is correct. Both technicians are correct. A bad axle bearing would build up excess heat, which could generate excess pressure and cause the axle seal to leak. The axle seals are not designed to hold back large quantities of axle fluid, so a overfilled differential could cause a leak at the axle seal.

Answer D is incorrect. Both technicians are correct.

27. A new universal joint has been installed in a vehicle. Technician A says the universal joint should be greased until grease purges out of all four caps. Technician B says that the universal joint is greased with a good quality lithium-based grease meeting *NLGI (national grease lubricating institute) Grade 1 or Grade 2*. Who is correct?

TASK D.2

    A. A only

    B. B only

    C. Both A and B

    D. Neither A nor B

Answer A is incorrect. Technician B is also correct.

Answer B is incorrect. Technician A is also correct.

Answer C is correct. Both technicians are correct. When installing universal joints, the new joints need to be lubricated with a good quality lithium-based grease meeting NLGI Grade 1 or 2 until it purges from all four caps.

Answer D is incorrect. Both technicians are correct.

28. Front universal joints are being replaced. Technician A says the new U-joints will still need to be greased after assembly. Technician B advises to pay close attention to the location of the zerk fitting—if installed incorrectly, then the zerks cannot be reached with a grease gun. Who is correct?

TASK F.7

    A. A only

    B. B only

    C. Both A and B

    D. Neither A nor B

Answer A is incorrect. Technician B is also correct.

Answer B is incorrect. Technician A is also correct.

Answer C is correct. Both technicians are correct. New U-joints need to be greased after installation until grease is purged from all four caps. Care must be taken to ensure proper location of the jerk fitting when installing U-joints.

Answer D is incorrect. Both technicians are correct.

TASK C.19

29. Fluid is being changed in a transaxle. Technician A says the proper fluid level is one finger joint through the fill hole. Technician B says most transaxles use 90-weight gear lube. Who is correct?

A. A only

B. B only

C. Both A and B

D. Neither A nor B

Answer A is incorrect. The proper fluid level for most transaxles is level with the bottom of the fill hole.

Answer B is incorrect. Most transaxles use motor oil or automatic transmission fluid instead of 90-weight gear oil.

Answer C is incorrect. Neither technician is correct.

Answer D is correct. Neither technician is correct.

TASK E.1.7

30. All of the following statements about differential case/carrier and ring-gear removal and replacement are true EXCEPT:

A. The ring-gear runout should be measured before removal of the case/carrier and ring-gear assembly.

B. The case/carrier side-play should be measured before removal of the case/carrier and ring-gear assembly.

C. The side bearing caps should be marked in relation to the housing before removal of the case/carrier and ring-gear assembly.

D. The side bearing should be clean and dry before installation of the case/carrier and ring-gear assembly.

Answer A is incorrect. The ring-gear runout should be checked to determine the condition of the trueness of the ring gear.

Answer B is incorrect. The case/carrier side-play should be measured before disassembly to determine the condition of the case/carrier.

Answer C is incorrect. Bearing caps should be marked and reinstalled in the same position that they were installed.

Answer D is correct. The side bearings should be lubricated before installation.

TASK B.5

31. A transmission is being disassembled for overhaul. Technician A says care should be taken to thoroughly inspect each component for damage as it is disassembled. Technician B says some transmission components have index marks that must be aligned when reassembled. Who is correct?

A. A only

B. B only

C. Both A and B

D. Neither A nor B

Answer A is incorrect. Technician B is also correct.

Answer B is incorrect. Technician A is also correct.

Answer C is correct. Both technicians are correct. During a transmission overhaul, all components should be thoroughly inspected, and any index marks parts should be checked for proper alignment.

Answer D is incorrect. Both technicians are correct.

32. When adjusting the pinion depth on the ring gear, technician A says to replace the collapsible spacer. Technician B says that pinion depth can also be adjusted by installing a selective pinion bearing race. Who is correct?

    TASK E.1.5

    A. A only

    B. B only

    C. Both A and B

    D. Neither A nor B

    Answer A is correct. Only technician A is correct. The collapsible spacer should always be replaced.

    Answer B is incorrect. Pinion depth is adjusted by installing shims under the pinion bearing.

    Answer C is incorrect. Only technician A is correct.

    Answer D is incorrect. Technician A is correct.

33. A transaxle has fluid leaking from the output shaft seal on an all-wheel drive vehicle. The technician removes the seal for inspection and cannot see any problem with the seal. Technician A says to check for a damaged yoke seal surface. Technician B says the vent could be plugged. Who is correct?

    TASK C.10

    A. A only

    B. B only

    C. Both A and B

    D. Neither A nor B

    Answer A is incorrect. Technician B is also correct.

    Answer B is incorrect. Technician A is also correct.

    Answer C is correct. Both technicians are correct. A seal leak can be caused due to many reasons, including a bad seal, a damaged yoke surface, or even a plugged vent, causing excessive pressures.

    Answer D is incorrect. Both technicians are correct.

34. A four-wheel drive vehicle has a jerking feel in the steering wheel when turning sharply with the four-wheel drive engaged. Technician A says this could be normal for many four-wheel drive vehicles. Technician B says if the vehicle has CV joints in the front axle and a center differential, then this would not be normal. Who is correct?

    TASK F.7

    A. A only

    B. B only

    C. Both A and B

    D. Neither A nor B

    Answer A is incorrect. Technician B is also correct.

    Answer B is incorrect. Technician A is also correct.

    Answer C is correct. Both technicians are correct. Many four-wheel drive vehicles use Cardan U-joints in the front outer axles; when turning sharply, they cause a jerking feeling in the steering wheel due to the angle. CV joints can be turned as sharp as 40 degrees without the jerking feel that Cardan U-joints cause.

    Answer D is incorrect. Both technicians are correct.

TASK B.11

35. In a manual 5-speed transmission, the reverse idler gear is in mesh with what other gear?

   A. First speed gear

   B. Fifth speed gear

   C. Main drive gear

   D. Reverse speed gear

Answer A is incorrect. The first speed gear is used for first gear and is in mesh with the first gear countershaft gear.

Answer B is incorrect. Fifth speed gear is used for fifth gear and is in mesh with the fifth gear countershaft gear.

Answer C is incorrect. The main drive gear is used for all gears except direct drive 1:1 and is in mesh with the main drive countershaft gear.

Answer D is correct. The reverse speed gear is in mesh with the reverse idler gear. The normal power flow in reverse is through the input shaft, through the main drive gear, to the main drive countershaft gear, through the countershaft to the reverse countershaft gear, through the reverse idler, to the reverse speed gear, and finally to the out shaft.

TASK F.8

36. Technician A says an unbalanced drive shaft causes a vibration that increases as vehicle speed increases. Technician B says an unbalanced drive shaft causes a vibration that decreases as vehicle speed increases. Who is correct?

   A. A only

   B. B only

   C. Both A and B

   D. Neither A nor B

Answer A is correct. Only technician A is correct. An unbalanced drive shaft causes a vibration that increases as vehicle speed increases.

Answer B is incorrect. An unbalanced drive shaft does not cause a vibration that decreases as vehicle speed increases.

Answer C is incorrect. Technician B is incorrect.

Answer D is incorrect. Technician A is correct.

TASK B.14

37. A vehicle has an inoperative speedometer. Technician A says the vehicle speed sensor could be defective. Technician B says the speedometer drive gear could be broken. Who is correct?

   A. A only

   B. B only

   C. Both A and B

   D. Neither A nor B

Answer A is incorrect. Technician B is also correct.

Answer B is incorrect. Technician A is also correct.

Answer C is correct. Both technicians are correct. An inoperative speedometer can be the result of a defective vehicle speed sensor or a broken speedometer drive gear.

Answer D is incorrect. Both technicians are correct.

38. Technician A says vehicle speed sensors can be tested with an ammeter. Technician B says when you check a vehicle speed sensor signal, you should use an oscilloscope while the vehicle is being driven. Who is correct?

TASK C.13

   A. A only
   B. B only
   C. Both A and B
   D. Neither A nor B

   Answer A is incorrect. A speed sensor can be a Hall-effect sensor or a magnetic pulse generator. They are not checked with an ammeter.

   Answer B is correct. Only technician B is correct. The best method to check speed sensors is by using an oscilloscope. A magnetic pulse generating speed sensor can be checked with an ohmmeter. An oscilloscope, however, will give the technician more information.

   Answer C is incorrect. Only technician B is correct.

   Answer D is incorrect. Technician B is correct.

39. A vacuum-shifted 4WD system does not shift into 4WD. Technician A says the vacuum hose that supplies the 4WD vacuum motor may be cracked. Technician B says the vacuum motor at the front axle may be the problem. Who is right?

TASK F.1

   A. A only
   B. B only
   C. Both A and B
   D. Neither A nor B

   Answer A is incorrect. Technician B is also correct.

   Answer B is incorrect. Technician A is also correct.

   Answer C is correct. Both technicians are correct. A cracked vacuum hose or a faulty vacuum motor at the front axle would cause no shift into 4WD.

   Answer D is incorrect. Both technicians are correct.

40. A vehicle is being diagnosed for a driveline vibration. Technician A says the drive shaft should be checked for proper phasing. Technician B says that drive shaft runout should be checked. Who is correct?

TASKS D2 & D5

   A. A only
   B. B only
   C. Both A and B
   D. Neither A nor B

   Answer A is incorrect. Technician B is also correct.

   Answer B is incorrect. Technician A is also correct.

   Answer C is correct. Both technicians are correct. A drive shaft out of phase could cause a driveline vibration. If the driveline is found to be in phase, however, then the drive shaft runout should be checked.

   Answer D is incorrect. Both technicians are correct.

PREPARATION EXAM 2—ANSWER KEY

1.	C	21.	C
2.	C	22.	C
3.	B	23.	A
4.	C	24.	A
5.	B	25.	C
6.	D	26.	B
7.	D	27.	A
8.	B	28.	C
9.	B	29.	B
10.	A	30.	D
11.	C	31.	C
12.	C	32.	C
13.	C	33.	C
14.	C	34.	C
15.	C	35.	C
16.	B	36.	B
17.	C	37.	A
18.	C	38.	C
19.	C	39.	C
20.	B	40.	B

PREPARATION EXAM 2—EXPLANATIONS

TASK A.5

1. Technician A says a close inspection of the transmission input shaft is part of a clutch replacement procedure. Technician B says the transmission input shaft can be replaced without completely disassembling the transmission on some transmissions. Who is correct?

   A. A only

   B. B only

   C. Both A and B

   D. Neither A nor B

   Answer A is incorrect. Technician B is also correct.

   Answer B is incorrect. Technician A is also correct.

   Answer C is correct. Both technicians are correct. The transmission input shaft splines and pilot should always be inspected during the installation of a new clutch.

   Answer D is incorrect. Both technicians are correct.

2. Ring-gear runout is being measured on a differential ring gear. Technician A says if ring-gear runout is excessive, then the ring gear can be moved 180 degrees on the ring-gear carrier and rechecked. Technician B says ring-gear runout should be checked before disassembly. Who is correct?

TASK E.1.3

 A. A only
 B. B only
 C. Both A and B
 D. Neither A nor B

Answer A is incorrect. Technician B is also correct.

Answer B is incorrect. Technician A is also correct.

Answer C is correct. Both technicians are correct. Ring-gear runout is measured using a dial indicator on the back side of the ring gear. Total indicated runout must be determined by adding all positive and negative readings. Measuring ring-gear runout before disassembly may assist in diagnosing rear-end complaints. If runout is found to be excessive, then the ring gear can be removed and remounted 180 degrees from its original location. If this does not correct the problem, then a new ring gear and pinion may be required.

Answer D is incorrect. Both technicians are correct.

3. A 5-speed transmission has been rebuilt because of a chipped tooth. The transmission now gets hung in two gears. Technician A says too stout of detent spring was installed. Technician B says the interlock should be inspected for proper operation. Who is correct?

TASK B.6

 A. A only
 B. B only
 C. Both A and B
 D. Neither A nor B

Answer A is incorrect. A detent spring is used to apply pressure to the detent. A detent spring that is too strong would cause hard shifting, but does not have anything to do with a transmission going into two gears at a time.

Answer B is correct. Only technician B is correct. The interlock prevents the transmission from being shifted into two gears at the same time. A faulty interlock or an interlock pin left out will allow the transmission to get hung in two gears.

Answer C is incorrect. Only technician B is correct.

Answer D is incorrect. Technician B is correct.

4. Technician A says the first step in transaxle removal is to disconnect the negative battery cable from the battery. Technician B says guide pins should be used to help support the transmission as it is slid out. Who is correct?

TASK C.4

 A. A only
 B. B only
 C. Both A and B
 D. Neither A nor B

Answer A is incorrect. Technician B is also correct.

Answer B is incorrect. Technician A is also correct.

Answer C is correct. Both technicians are correct. The very first thing the technician should do is disconnect the negative battery cable from the battery. After the negative battery cable is disconnected, the technician can begin disconnecting other connections such as clutch linkage, shift cables, back-up light wiring, and any other component connections. The technicians should install two guide studs made from cap screws with the head cut off. The guide studs will help support the weight of the transmission as it slides out.

Answer D is incorrect. Both technicians are correct.

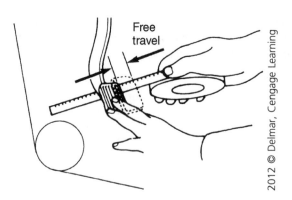

Free
travel

2012 © Delmar, Cengage Learning

TASK A.2

5. Technician A says the clutch safety switch free travel is being measured in the illustration. Technician B says that as the clutch disc wears, this free travel will decrease. Who is correct?

    A. A only

    B. B only

    C. Both A and B

    D. Neither A nor B

    Answer A is incorrect. While some clutch safety switches are adjustable, the free travel is not a consideration with a clutch switch.

    Answer B is correct. Only technician B is correct. The clutch pedal free travel is the measurement of the distance the release bearing travels before making contact with the pressure plate release finger. As the clutch disc wears, the free travel will decrease.

    Answer C is incorrect. Only technician B is correct.

    Answer D is incorrect. Technician B is correct.

TASK D.2

6. Technician A says that the removal of a drive shaft usually requires partial disassembly of the front suspension. Technician B says the first step in drive shaft removal is to loosen the rear differential yoke nut. Who is correct?

    A. A only

    B. B only

    C. Both A and B

    D. Neither A nor B

    Answer A is incorrect. The front suspension does not have to be disassembled in order to remove a drive shaft.

    Answer B is incorrect. The rear differential yoke nut cannot be removed until the drive shaft is removed.

    Answer C is incorrect. Neither technician is correct.

    Answer D is correct. Neither technician is correct. The front suspension does not have to be disassembled in order to remove a drive shaft. The rear differential yoke nut cannot be removed until the drive shaft is removed.

7. Technician A says the collapsible pinion shaft spacer may be reused if the differential is disassembled and overhauled. Technician B says that after proper pinion bearing preload is set, pinion depth is adjusted by backing off the pinion flange nut. Who is right?

TASK E.1.4

   A. A only
   B. B only
   C. Both A and B
   D. Neither A nor B

   Answer A is incorrect A collapsible pinion shaft spacer is not reusable and should be discarded after disassembly.

   Answer B is incorrect. No pinion adjustment is made by loosening the nut once proper bearing preload has been established.

   Answer C is incorrect. Neither technician is correct.

   Answer D is correct. Neither technician is correct. A collapsible pinion shaft spacer is not reusable and should be discarded after disassembly. No pinion adjustment is made by loosening the nut once proper bearing preload has been established.

8. An all-wheel drive vehicle has a vibration that is more noticeable while cornering. Technician A says the U-joints in the drive shaft may be worn. Technician B says the outboard front axle joints may be worn. Who is right?

TASK F.1

   A. A only
   B. B only
   C. Both A and B
   D. Neither A nor B

   Answer A is incorrect. If the vehicle's U-joints were worn to the point of causing a vibration, then it would not only be evident while cornering, but at all speeds.

   Answer B is correct. Only technician B is correct. Outboard U-joint vibration increases during a turn.

   Answer C is incorrect. Only technician B is correct.

   Answer D is incorrect. Technician B is correct.

9. The counter/cluster gear shaft and needle bearings in a 4-speed transmission are pitted and scored. Technician A says the transmission may have a growling noise with the engine idling, transmission in neutral, and the clutch pedal out. Technician B says the transmission may have a growling noise while driving in any gear except fourth gear. Who is right?

TASK B.1

   A. A only
   B. B only
   C. Both A and B
   D. Neither A nor B

   Answer A is incorrect. The counter/cluster shaft bearings are not loaded at idle and would not make noise.

   Answer B is correct. The counter/cluster shaft bearings are under load any time the transmission is operating in gear with the exception of fourth, because fourth gear is direct drive.

   Answer C is incorrect. Only technician B is correct.

   Answer D is incorrect. Technician B is correct.

TASK D.2

10. A front-wheel drive vehicle has a clicking noise while turning. Technician A says this could be caused by a worn outer-drive axle joint. Technician B says this could be caused by a worn front wheel bearing. Who is right?

A. A only

B. B only

C. Both A and B

D. Neither A nor B

Answer A is correct. The change in the turning angles inside the constant-velocity, outer-drive axle joint during a turn will result in a noise that is not present when driving in a straight line.

Answer B is incorrect. A wheel bearing usually makes a grinding or growling noise that may change during turns.

Answer C is incorrect. Only technician A is correct.

Answer D is incorrect. Technician A is correct.

TASK D.3

11. A rear-wheel drive (RWD) vehicle has a clunking noise when accelerating from a stop. Technician A says if this vehicle uses a center-support bearing, then it should be inspected for proper operation. Technician B says the center-support bearing can be shimmed to adjust driveline angle. Who is correct?

A. A only

B. B only

C. Both A and B

D. Neither A nor B

Answer A is incorrect. Technician B is also correct.

Answer B is incorrect. Technician A is also correct.

Answer C is correct. Both technicians are correct. On vehicles with long wheel bases and some high-performance vehicles, it is necessary to use a center-support bearing to control drive shaft vibration, allowing two short drive shafts instead of one long drive shaft. If the universal joint working angles are found to be incorrect, then shimming can be used on the center-support bearing bracket to adjust the working joint angle.

Answer D is incorrect. Both technicians are correct.

TASK C.1

12. A transaxle is being diagnosed for hard shifting. Technician A says the lubricant should be checked for proper viscosity. Technician B says if a highly viscous oil is used, then the hard shifting will be more noticeable in colder weather. Who is correct?

A. A only

B. B only

C. Both A and B

D. Neither A nor B

Answer A is incorrect. Technician B is also correct.

Answer B is incorrect. Technician A is also correct.

Answer C is correct. Both technicians are correct. Any time a shifting concern is being diagnosed, it is good practice to verify that proper lubricating oil has been used. Many cars require automatic transmission fluid, while others require a multigrade motor oil. If standard GL-5 gear oil is used for these vehicles, then they will have hard or stiff shifting, especially in cold weather.

Answer D is incorrect. Both technicians are correct.

13. A four-wheel drive with a manual shifting linkage is being adjusted. Technician A says some transfer cases require the 4WD lever be placed in a specific position before adjustments can be made. Technician B says some transfer cases require the use of a spacer of a certain size to position the lever properly for adjustment. Who is correct?

TASK F.2

   A. A only

   B. B only

   C. Both A and B

   D. Neither A nor B

   Answer A is incorrect. Technician B is also correct.

   Answer B is incorrect. Technician A is also correct.

   Answer C is correct. Both technicians are correct. Most manually shifted transfer cases have provisions for adjusting the shift linkage. The actual adjustment will vary between makes and models—some require the lever be in a certain position, while others require the use of a specific size shift lever spacer.

   Answer D is incorrect. Both technicians are correct.

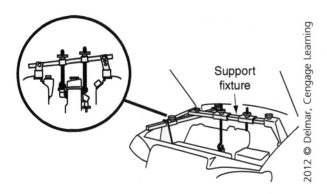

Support fixture

2012 © Delmar, Cengage Learning

14. Technician A says the transaxle shown in the illustration is being removed and the support fixture is used to support the engine during transaxle removal. Technician B says the support fixture is used to help position the engine during transaxle installation. Who is correct?

TASK A.4

   A. A only

   B. B only

   C. Both A and B

   D. Neither A nor B

   Answer A is incorrect. Technician B is also correct.

   Answer B is incorrect. Technician A is also correct.

   Answer C is correct. Both technicians are correct. On most front-wheel drive vehicles, the engine must be supported using a support fixture when the transaxle is removed. Most support fixtures provide a means to raise or lower the engine to aid in transaxle installation.

   Answer D is incorrect. Both technicians are correct.

TASK A.7

15. During a clutch replacement job, the flywheel is found to have too much runout. Technician A says sometimes the flywheel runout can be adjusted by re-torquing the flywheel. Technician B says excessive flywheel runout can cause poor release symptoms. Who is correct?

  A. A only

  B. B only

  C. Both A and B

  D. Neither A nor B

Answer A is incorrect. Technician B is also correct.

Answer B is incorrect. Technician A is also correct.

Answer C is correct. Both technicians are correct. Excessive flywheel runout can cause clutch symptoms such as no release, gear clash, grabbing and chatter. If flywheel runout is found to be out of limits, then the runout might be adjusted by re-torquing the flywheel. Loosen all of the flywheel bolts, then tighten the bolt closest to the high spot, and then tighten one of the bolts next to the first one. The third bolt to tighten is the other one next to the first bolt tightened. Continue this until the last bolt tightened is the one closest to the low spot.

Answer D is incorrect. Both technicians are correct.

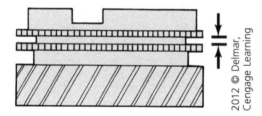

2012 © Delmar, Cengage Learning

TASK B.9

16. In reference to the illustration, Technician A says that as the synchronizer blocker ring wears, the distance between the two arrows will increase. Technician B says this distance is measured with a feeler gauge. Who is correct?

  A. A only

  B. B only

  C. Both A and B

  D. Neither A nor B

Answer A is incorrect. As the blocker ring wears, the distance will decrease.

Answer B is correct. Only technician B is correct. When inspecting synchronizers and blocker rings, the distance between the speed gear clutch teeth and the blocker ring is measured using a feeler gauge.

Answer C is incorrect. Only technician B is correct.

Answer D is incorrect. Technician B is correct.

17. A four-wheel drive vehicle has had a premature failure of the viscous coupler. Technician A says the tires should be checked for correct brand type and size. Technician B says that low pressure, as little as 5 psi, can effect operation of the viscous coupler. Who is correct?

TASK F.11

    A. A only

    B. B only

    C. Both A and B

    D. Neither A nor B

Answer A is incorrect. Technician B is also correct.

Answer B is incorrect. Technician A is also correct.

Answer C is correct. Both technicians are correct. The viscous coupler allows the front and rear axle to rotate at different speeds. Tire size, brand, and type are crucial for proper operation of the four-wheel drive system. As little as 5 psi, or unequal tire diameters ¼" or more, can effect operation.

Answer D is incorrect. Both technicians are correct.

18. Technician A says a plugged transfer case vent can cause fluid leakage from a seal. Technician B says that a remotely located transfer case vent helps to prevent water from entering the transfer case when driving through high waters. Who is right?

TASK F.4

    A. A only

    B. B only

    C. Both A and B

    D. Neither A nor B

Answer A is incorrect. Technician B is also correct.

Answer B is incorrect. Technician A is also correct.

Answer C is correct. Both technicians are correct. The case vent allows excess pressure to vent from the case as the fluid temperature increases. If the vent gets plugged, then the excess pressure can push fluid past the seals. Many 4WD vehicles incorporate a remotely located vent at a much higher level than the transfer case to prevent water from entering the transfer case when driving through high waters.

Answer D is incorrect. Both technicians are correct.

19. A rear-wheel drive vehicle has a vibration that increases in relation to vehicle speed. Technician A says the counter weight may have fallen off the drive shaft. Technician B says some of the wheels and tires may be out of balance. Who is right?

TASK D.4

    A. A only

    B. B only

    C. Both A and B

    D. Neither A nor B

Answer A is incorrect. Technician B is also correct.

Answer B is incorrect. Technician A is also correct.

Answer C is correct. Both technicians are correct. A vehicle that vibrates in relationship to speed is typically due to driveline or tire balance. Most vibrations that start at 60 mph are due to balance issues, whether tire or drive shaft.

Answer D is incorrect. Both technicians are correct.

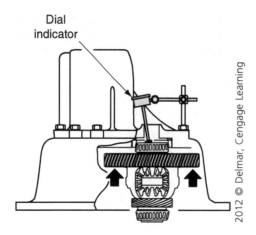

Dial indicator

2012 © Delmar, Cengage Learning

TASK C.17

20. The measurement in the illustration determines the proper

   A. Bearing wear
   B. Side bearing preload
   C. Side gear end-play
   D. Bearing race depth

   Answer A is incorrect. Bearing wear is examined by feel and vision.

   Answer B is correct. A dial indicator is used to determine which size shim to install in order to adjust the differential side-bearing preload.

   Answer C is incorrect. The dial indicator is not positioned in a way side gear end-play can be measured.

   Answer D is incorrect. The dial indicator is not positioned in a way bearing race depth can be measured.

TASK F.1

21. Technician A says the primary difference between full-time four-wheel drive and part-time four-wheel drive is the center differential. Technician B says the four-wheel drive on a vehicle with part-time four-wheel drive is for off-road only. Who is correct?

   A. A only
   B. B only
   C. Both A and B
   D. Neither A nor B

   Answer A is incorrect. Technician B is also correct.

   Answer B is incorrect. Technician A is also correct.

   Answer C is correct. Both technicians are correct. There are two major types of four-wheel drive vehicles. There is part-time four-wheel drive and full-time four-wheel drive, also called *all-wheel drive* (AWD). A vehicle with part-time 4WD should not be engaged in 4WD unless being used for off-road use. Part-time 4WD has positive, mechanical connections between the front and rear axle, unlike the full-time 4WD, which has a center differential.

   Answer D is incorrect. Both technicians are correct.

22. A vehicle has an oil-contaminated clutch disc. Technician A says the source of the oil could be the rear main seal in the engine. Technician B says the source of the oil could be from the engine valve covers. Who is correct?

TASK A.8

A. A only

B. B only

C. Both A and B

D. Neither A nor B

Answer A is incorrect. Technician B is also correct.

Answer B is incorrect. Technician A is also correct.

Answer C is correct. Both technicians are correct. A leaking rear main seal or leaking valve covers could both cause the clutch disc to become oil-contaminated.

Answer D is incorrect. Both technicians are correct.

23. Technician A says misadjusted shift linkage can cause the transmission to jump out of gear. Technician B says misadjusted shift linkage can cause the transmission to have hard shifting. Who is correct?

TASK B.2

A. A only

B. B only

C. Both A and B

D. Neither A nor B

Answer A is correct. Only technician A is correct. Shift linkage adjustment is only possible on a transmission with external shift linkage. If the linkage is not adjusted properly, then the transmission can jump out of the selected gear, usually under a load.

Answer B is incorrect. A hard-shifting complaint is usually associated with binding shift rails, incorrect lubricant, or even clutch linkage/adjustment problems.

Answer C is incorrect. Only technician A is correct.

Answer D is incorrect. Technician A is correct.

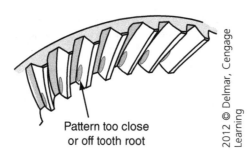

Pattern too close
or off tooth root

2012 © Delmar, Cengage
Learning

TASK E.1.8

24. Technician A says the pinion depth shown in the illustration is too deep. Technician B says ring-gear backlash should be decreased. Who is correct?

    A. A only
    B. B only
    C. Both A and B
    D. Neither A nor B

    Answer A is correct. Only technician A is correct. The tooth contact pattern shown is too deep and too close to the toe of the tooth. If increasing pinion tooth depth, then the pinion depth also would need to be increased. If the pattern is too close to the toe of the tooth, then the ring gear needs to be moved away from the pinion, which would increase backlash.

    Answer B is incorrect. The backlash needs to be increased to move the ring gear away from the pinion.

    Answer C is incorrect. Only technician A is correct.

    Answer D is incorrect. Technician A is correct.

TASK C.5

25. A transaxle is being disassembled for overhaul. Technician A says the input and main shaft end-play should be measured before disassembly. Technician B says the input shaft should be rotated and the transaxle shifted through the gears including reverse gear. Who is correct?

    A. A only
    B. B only
    C. Both A and B
    D. Neither A nor B

    Answer A is incorrect. Technician B is also correct.

    Answer B is incorrect. Technician A is also correct.

    Answer C is correct. Both technicians are correct. Always measure end-play of the input and mainshaft before disassembly. This information will be needed during the reassembly. Rotating the input shaft while shifting through the gears can help identify problems before disassembly.

    Answer D is incorrect. Both technicians are correct.

26. Technician A says the flywheel can be shimmed if the crankshaft end-play is found to be excessive. Technician B says the flywheel runout should be checked every time the clutch is replaced. Who is correct?

    A. A only

    B. B only

    C. Both A and B

    D. Neither A nor B

TASK A.9

Answer A is incorrect. If the crankshaft end-play is found to be excessive, then the crankshaft thrust washer must be replaced.

Answer B is correct. Only technician B is correct. It is always good practice to measure the flywheel runout when the clutch is being replaced. If machining is necessary, then shims might need to be added between the flywheel and crank to compensate for the removal of material from the flywheel.

Answer C is incorrect. Only technician B is correct.

Answer D is incorrect. Technician B is correct.

27. Technician A says a limited-slip differential can be checked using a torque wrench. Technician B says a limited-slip differential can be checked using a spring scale. Who is correct?

    A. A only

    B. B only

    C. Both A and B

    D. Neither A nor B

TASK E.3.1

Answer A is correct. Only technician A is correct. The breaking torque of a limited-slip differential is checked by jacking one wheel off the ground and measuring the amount of foot pounds it takes to rotate the jacked-up wheel.

Answer B is incorrect. A spring scale is usually used to measure pinion bearing preload, not the operation of limited slip differential.

Answer C is incorrect. Only technician A is correct.

Answer D is incorrect. Technician A is correct.

28. An axle shaft on a rear-wheel drive rear end has excessive end-play. Technician A says the problem could be a faulty axle bearing. Technician B says the problem could be a worn C-lock retainer. Who is correct?

    A. A only

    B. B only

    C. Both A and B

    D. Neither A nor B

TASK E.4.4

Answer A is incorrect. Technician B is also correct.

Answer B is incorrect. Technician A is also correct.

Answer C is correct. Both technicians are correct. When measuring axle shaft end-play, normal readings are 0.005–0.030 inches (0.13–0.7 mm). Excessive readings indicate faulty axle bearing or too thin a C-lock.

Answer D is incorrect. Both technicians are correct.

TASK F.12

29. Technician A says the tire speed index has no effect on all-wheel drive operation. Technician B says mixing tire brands can have an effect on all-wheel drive vehicles. Who is correct?

A. A only

B. B only

C. Both A and B

D. Neither A nor B

Answer A is incorrect. The speed index is used for rating how fast a tire can be used safely.

Answer B is correct. Only technician B is correct. When considering all-wheel drive, the tire brand can matter, because it is crucial that all of the tires be the same size. Mismatching with different brand tires can cause the center differential to overwork due to tire size variations among brands.

Answer C is incorrect. Only technician B is correct.

Answer D is incorrect. Technician B is correct.

TASK B.3

30. Technician A says it is acceptable to reuse lock washers when overhauling a transmission. Technician B says anaerobic sealers cure in the presence of air. Who is correct?

A. A only

B. B only

C. Both A and B

D. Neither A nor B

Answer A is incorrect. It is never ok to reuse lock washers or lock nuts.

Answer B is incorrect. Anaerobic sealers cure in the absence of air. (RTV) room temperature vulcanizing sealers cure in the presence of air.

Answer C is incorrect. Neither technician is correct.

Answer D is correct. Neither technician is correct. Lock washers should never be reused, and anaerobic sealers cure in the absence of air.

TASK B.4

31. Technician A says guide studs should be used when removing a transmission. Technician B says guide studs should be used when installing a transmission. Who is correct?

A. A only

B. B only

C. Both A and B

D. Neither A nor B

Answer A is incorrect. Technician B is also correct.

Answer B is incorrect. Technician A is also correct.

Answer C is correct. Both technicians are correct. When removing and installing a transmission, guide studs should be used to help support the weight of the transmission.

Answer D is incorrect. Both technicians are correct.

32. Technician A says if a spreader is being used to remove a differential from the housing assembly, then the housing should not be spread more than 0.030 inches (0.762 mm). Technician B says if a spreader is not available, then the differential can be pried out using a pry bar and a piece of wood. Who is correct?

TASK E.2.2

    A. A only
    B. B only
    C. Both A and B
    D. Neither A nor B

Answer A is incorrect. The maximum distance an axle housing can be spread without damage is 0.015 inches (0.381mm).

Answer B is correct. Only technician B is correct. If a spreader is not available, then the differential can be pried out using a pry bar and a piece of wood to protect the sealing surface of the axle housing. Another method is to use a box end wrench on one of the ring gear bolts and rotate the pinion.

Answer C is incorrect. Only technician B is correct.

Answer D is incorrect. Technician B is correct.

33. Technician A says if the transfer-case shaft assembly end-play is excessive, then selective snap rings may be used to get the end-play within tolerances. Technician B says some transfer cases use shims to adjust transfer-case shaft assembly end-play. Who is correct?

TASK F.5

    A. A only
    B. B only
    C. Both A and B
    D. Neither A nor B

Answer A is incorrect. Technician B is also correct.

Answer B is incorrect. Technician A is also correct.

Answer C is correct. Both technicians are correct. When reassembling the transfer case, the shaft assembly end-play should be measured using a dial indicator. If the end-play is out of tolerances, then selective snap ring and shims can be installed to get the end-play within an acceptable tolerance range.

Answer D is incorrect. Both technicians are correct.

34. A speedometer is inoperative on a front-wheel drive vehicle. Technician A says the vehicle speed sensor could be the problem and should be checked. Technician B says a scan tool can be used to verify that the vehicle speed sensor (VSS) is operating correctly. Who is correct?

TASK C.11

    A. A only
    B. B only
    C. Both A and B
    D. Neither A nor B

Answer A is incorrect. Technician B is also correct.

Answer B is incorrect. Technician A is also correct.

Answer C is correct. Both technicians are correct. The VSS is responsible for speed input to the powertrain control module or the body control module. A scan tool can be used to see if the VSS is actually sending out a signal. The VSS is driven by the transaxle. If the VSS is not sending a signal. then it should be removed and the speedometer drive gear in the transaxle inspected.

Answer D is incorrect. Both technicians are correct.

TASK B.4

35. Technician A says the drive shaft should be marked before removing it during transmission removal. Technician B says some transmissions must be removed as a unit with the engine. Who is correct?

   A.   A only
   B.   B only
   C.   Both A and B
   D.   Neither A nor B

   Answer A is incorrect. Technician B is also correct.

   Answer B is incorrect. Technician A is also correct.

   Answer C is correct. Both technicians are correct. The drive shaft should always be marked when removing it to ensure proper phasing when reinstalling. On some vehicles, the engine and transmission must be removed as a unit.

   Answer D is incorrect. Both technicians are correct.

TASK D.1

36. A front-wheel drive vehicle has a vibration in the steering wheel at highway speeds. Technician A says the inner CV joint could be worn and not operating smoothly. Technician B says the front wheels may be out of balance. Who is correct?

   A.   A only
   B.   B only
   C.   Both A and B
   D.   Neither A nor B

   Answer A is incorrect. A vibration felt in the steering wheel at highway speeds is an indication that the front wheels are out of balance. A worn inner CV joint would vibrate throughout the entire vehicle, not just though the steering wheel.

   Answer B is correct. Only technician B is correct. This is a classic example of a vehicle with the front wheels out of balance.

   Answer C is incorrect. Only technician B is correct.

   Answer D is incorrect. Technician B is correct.

TASK E.2.1

37. A rear-wheel drive vehicle has a growling noise coming from the rear axle that is only noticed in turns. Technician A says the problem is in the differential. Technician B says it could be due to worn carrier bearings. Who is correct?

   A.   A only
   B.   B only
   C.   Both A and B
   D.   Neither A nor B

   Answer A is correct. Only technician A is correct. A noise in the rear axle that gets louder in turns is in the differential. When going straight, the differential is not operating unless the vehicle has mismatched tires. The differential allows one wheel to operate at a different speed than the other during turns.

   Answer B is incorrect. Worn carrier bearings produce a low-pitch rumble at more than 20 mph.

   Answer C is incorrect. Only technician A is correct.

   Answer D is incorrect. Technician A is correct.

38. Technician A says the transaxle must be disassembled in order to replace the differential on a FWD vehicle. Technician B says most transaxles are lubricated by gear rotation and troughs and funnels. Who is correct?

TASKS C.14
& C18

   A. A only
   B. B only
   C. Both A and B
   D. Neither A nor B

   Answer A is incorrect. Technician B is also correct.

   Answer B is incorrect. Technician A is also correct.

   Answer C is correct. Both technicians are correct. The differential on a transaxle is an integral part of the transaxle. To replace the differential, the transaxle must be disassembled. Like transmissions, transaxles are lubricated by the oil passing over the rotating gears and directing the oil to critical areas using troughs and oiling funnels.

   Answer D is incorrect. Both technicians are correct.

39. Technician A says transfer cases are lubricated with automatic transmission fluid (ATF). Technician B says transfer cases are lubricated with gear oil. Who is correct?

TASK F.6

   A. A only
   B. B only
   C. Both A and B
   D. Neither A nor B

   Answer A is incorrect. Technician B is also correct.

   Answer B is incorrect. Technician A is also correct.

   Answer C is correct. Both technicians are correct. Some transfer cases use ATF and others use gear oil. The correct level is even with the bottom of the fill plug.

   Answer D is incorrect. Both technicians are correct.

40. Technician A says the side-bearing preload on a transaxle can be set using threaded adjusters. Technician B says the side-bearing preload on a transaxle can be set using shims. Who is correct?

TASK C.15

   A. A only
   B. B only
   C. Both A and B
   D. Neither A nor B

   Answer A is incorrect. Transaxles do not have the adjustable end plugs like some rear ends use. They use shims to adjust the side-bearing preload.

   Answer B is correct. Only technician B is correct. The side-bearing preload is adjusted using shims. If a preload is specified, then the shim size needs to be size of the amount of clearance plus an additional 0.003 inches (0.07mm).

   Answer C is incorrect. Only technician B is correct.

   Answer D is incorrect. Technician B is correct.

PREPARATION EXAM 3—ANSWER KEY

1.	C	21.	C
2.	A	22.	B
3.	C	23.	C
4.	C	24.	C
5.	B	25.	B
6.	D	26.	C
7.	A	27.	B
8.	C	28.	B
9.	A	29.	A
10.	C	30.	C
11.	A	31.	C
12.	D	32.	C
13.	C	33.	C
14.	B	34.	C
15.	B	35.	D
16.	C	36.	A
17.	C	37.	A
18.	B	38.	D
19.	C	39.	A
20.	B	40.	C

PREPARATION EXAM 3—EXPLANATIONS

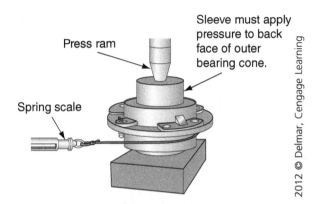

Press ram

Sleeve must apply pressure to back face of outer bearing cone.

Spring scale

2012 © Delmar, Cengage Learning

1. Technician A says pinion bearing preload is being set in the illustration. Technician B says the press simulates the clamping force of the yoke without having to actually install the yoke. Who is correct?

    TASK E.1.6

    A. A only
    B. B only
    C. Both A and B
    D. Neither A nor B

    Answer A is incorrect. Technician B is also correct.

    Answer B is incorrect. Technician A is also correct.

    Answer C is correct. Both technicians are correct. Pinion bearing preload is adjusted with a collapsible spacer or shims. On some remote differentials the clamping force of the yoke can be simulated using a hydraulic press with a predetermined amount of force by the manufacturer. Preload can be measured in a couple of different ways. One way is with an inch pound torque wrench measuring the rotating resistance. The other way is with a spring scale, as seen in the figure, to also measure rotating resistance.

    Answer D is incorrect. Both technicians are correct.

2. Clutch housing bore runout is excessive on a RWD vehicle. Technician A says the symptom for excessive clutch housing bore runout is premature pilot bearing/bushing wear. Technician B says the maximum general specification for clutch housing bore runout is 0.020 inches (0.50 mm). Who is correct?

    TASK A.8

    A. A only
    B. B only
    C. Both A and B
    D. Neither A nor B

    Answer A is correct. Only technician A is correct. Excessive clutch housing bore runout can cause chatter, transmission jumping out of gear, grabbing, no release and premature pilot bearing/bushing wear.

    Answer B is incorrect. The maximum allowable clutch housing bore runout is 0.010 inches (0.254 mm).

    Answer C is incorrect. Only technician A is correct.

    Answer D is incorrect. Technician A is correct.

TASK C.9

3. Technician A says the idler gear is used for reverse gear only. Technician B says the idler gear end-play is controlled by thrust washers. Who is correct?

   A. A only
   B. B only
   C. Both A and B
   D. Neither A nor B

   Answer A is incorrect. Technician B is also correct.

   Answer B is incorrect. Technician A is also correct.

   Answer C is correct. Both technicians are correct. The idler gear in a transmission or transaxle is used to reverse the power flow for reverse. The reverse idler gear is mounted on the reverse idler shaft and uses thrust washers to control end-play.

   Answer D is incorrect. Both technicians are correct.

TASK B.17

4. Technician A advises to ensure the vehicle is level when checking the transmission fluid. Technician B says changing the fluid is important to remove metal contaminants that can cause an increase of wear on internal transmission components. Who is correct?

   A. A only
   B. B only
   C. Both A and B
   D. Neither A nor B

   Answer A is incorrect. Technician B is also correct.

   Answer B is incorrect. Technician A is also correct.

   Answer C is correct. Both technicians are correct. Oil in a transmission cleans, cools, and lubricates the transmission. Some manufacturers recommend changing the lubricant after the first 5,000 miles (8,000 km). Always be sure the vehicle is level when checking the transmission fluid.

   Answer D is incorrect. Both technicians are correct.

TASK D.5

5. Technician A says that drive shaft runout should be measured 1 inch from the seam of the yoke. Technician B says the drive shaft runout should be measured at the center of the drive shaft. Who is correct?

   A. A only
   B. B only
   C. Both A and B
   D. Neither A nor B

   Answer A is incorrect. The runout on a drive shaft should be measured 3 inches from each end and at the center. The maximum allowable runout on the ends is 0.020 inches (0.508 mm) and 0.025 inches (.0635 mm) at the center.

   Answer B is correct. Only technician B is correct. The drive shaft runout should be measured at the end and at the center.

   Answer C is incorrect. Only technician B is correct.

   Answer D is incorrect. Technician B is correct.

6. All of the following are correct procedures when servicing wheel bearing EXCEPT:

   A. Ensure an end-play of 0.001–0.005 inches (0.0254–0.127 mm).

   B. The wheel bearing is cleaned in solvent and repacked.

   C. A new grease seal is installed.

   D. Use compressed air to spin-dry the bearing.

TASK F.9

Answer A is incorrect. The wheel bearing end-play should be adjusted to 0.001–0.005 inches (0.0254–0.127 mm).

Answer B is incorrect. The wheel bearing should be cleaned in solvent, dried, then repacked with clean, good quality grease.

Answer C is incorrect. A new grease seal should always be used when servicing wheel bearings.

Answer D is correct. It is never acceptable to spin-dry a wheel bearing. Compressed air can be used to dry the bearing. Do not allow the bearing to spin from the compressed air; serious bodily harm can occur from this practice.

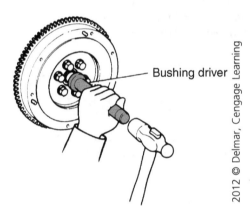

Bushing driver

2012 © Delmar, Cengage Learning

7. What is the technician doing in the above illustration?

   A. Installing a pilot bearing

   B. Removing a pilot bearing

   C. Installing a release bearing

   D. Removing a release bearing

TASK A.6

Answer A is correct. The technician is installing a pilot bearing.

Answer B is incorrect. The technician is installing a pilot bearing.

Answer C is incorrect. The technician is installing a pilot bearing. A release bearing would be installed in the clutch housing, on the transmission input bearing retainer.

Answer D is incorrect. The technician is installing a pilot bearing. A release bearing would be installed in the clutch housing, on the transmission input bearing retainer.

TASK C.6

8. A transaxle is jumping out of third gear when driving on a rough terrain. Technician A says the detent spring could be broken or weak. Technician B says the detent could be worn. Who is correct?

   A. A only

   B. B only

   C. Both A and B

   D. Neither A nor B

   Answer A is incorrect. Technician B is also correct.

   Answer B is incorrect. Technician A is also correct.

   Answer C is correct. Both technicians are correct. The detent and detent spring are designed to help hold the transaxle in the selected gear. If the detent gets worn or the detent spring is broken or gets weak, then the transmission can jump out of gear, especially when driving on rough roads.

   Answer D is incorrect. Both technicians are correct.

TASK E.1.2

9. A drive axle has repeated pinion seal failure due to leaks. Technician A says the sealing surface of the seal should be inspected for a wear groove. Technician B says the collapsible spacer may need replacing to fix the seal failure complaint. Who is correct?

   A. A only

   B. B only

   C. Both A and B

   D. Neither A nor B

   Answer A is correct. Only technician A is correct. Repeat pinion seal failure due to leakage could be the result of a yoke with a seal wear groove, which allows the fluid leak. If this is the case, then the yoke should be replaced.

   Answer B is incorrect. The collapsible spacer is used to set pinion bearing preload and should be replaced each time the pinion yoke is removed. It does not affect seal performance.

   Answer C is incorrect. Only technician A is correct.

   Answer D is incorrect. Technician A is correct.

TASK B.12

10. Excessive input shaft end-play in a 4-speed transmission may cause the transmission to jump out of:

   A. First gear

   B. Second gear

   C. Fourth gear

   D. Reverse gear

   Answer A is incorrect. First gear is at the rear of the transmission, away from the input shaft. The power flow travels in the input shaft down through the countershaft, up through the speed gear, then out the output shaft.

   Answer B is incorrect. Second gear is the second closest gear to the rear of the transmission. The power flow travels in the input shaft down through the countershaft, up through the speed gear, then out the output shaft.

   Answer C is correct. Fourth gear is the closest to the input shaft. Excessive end-play would allow too much movement of the input shaft. In fourth gear, the input shaft is splined to the main shaft in direct line.

   Answer D is incorrect. Reverse gear is at the rear of the transmission. The power flow is in the input shaft, down the countershaft to the reverse idler gear, then to the reverse speed gear and out the output shaft.

11. A front-wheel drive car has a clicking noise in the front while turning. Technician A says this could be caused by a worn outer-drive axle joint. Technician B says this could be caused by a front wheel bearing. Who is correct?

TASK D.2

   A. A only
   B. B only
   C. Both A and B
   D. Neither A nor B

   Answer A is correct. The change in the position and required movement of the bearings inside an outer drive axle joint during a turn will result in a noise that is not present when driving in a straight line; this noise is usually a clicking noise.

   Answer B is incorrect. A wheel bearing usually makes a grinding or growling noise that may change during turns.

   Answer C is incorrect. Only technician A is correct.

   Answer D is incorrect. Technician A is correct.

12. All of the following are true about transfer-case disassembly and inspection and assembly EXCEPT:

   A. The transfer-case chain should be inspected for stretching and looseness.
   B. Thrust washer thickness should be measured to check for wear or for proper select fit.

   TASK F.2

   C. Specification measurements should have been taken during disassembly.
   D. All parts should be cleaned before installation and assembled dry.

   Answer A is incorrect. Over time, the transfer case chain stretches. The chain should always be inspected for any wear or stretch and replaced accordingly.

   Answer B is incorrect. Select-fit thrust washers should be measured and used for assembly of the transfer case to adjust end-play to the proper specification. Select fit means it comes in various thicknesses.

   Answer C is incorrect. During disassembly, the transfer case components should be thoroughly inspected to ensure any worn or questionable parts are replaced during the rebuild.

   Answer D is correct. All of the transfer case parts should be cleaned then lubricated before assembly. Any time bearings are cleaned and inspected, it is a good idea to lubricate them and wrap them in an oil or wax paper to prevent rust until used.

13. Technician A says the marcel spring allows smooth clutch engagement and longer drive train life. Technician B says the dampening springs compress during clutch engagement. Who is correct?

TASK A.1

   A. A only
   B. B only
   C. Both A and b
   D. Neither A nor B

   Answer A is incorrect. Technician B is also correct.

   Answer B is incorrect. Technician A is also correct.

   Answer C is correct. Both technicians are correct. A typical clutch disc has two different types of springs. The marcel spring, which is between the two clutch facings, allows smooth engagement. A failure in the marcel spring would result in harsh clutch engagement. The dampening spring or torsion springs absorb engine torsional vibration as the clutch is engaged.

   Answer D is incorrect. Both technicians are correct.

TASK B.7

14. Technician A says if the transmission input shaft shows wear, then only on the side of the spines it can be reused. Technician B says the front pilot area of the input shaft should be inspected. Who is correct?

A. A only

B. B only

C. Both A and B

D. Neither A nor B

Answer A is incorrect. Any wear on the input shaft spines can allow the clutch disc to hang up, causing poor clutch release

Answer B is correct. Only technician B is correct. The front pilot area of the input shaft is supported by the pilot bearing and should always be inspected for excessive wear.

Answer C is incorrect. Only technician B is correct.

Answer D is incorrect. Technician B is correct.

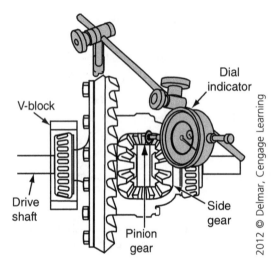

TASK C.15

15. Technician A says if the measurement being taken in the illustration is excessive, then the pinion gear thrust washer must be replaced. Technician B says this measurement can determine gear wear. Who is correct?

A. A only

B. B only

C. Both A and B

D. Neither A nor B

Answer A is incorrect. Side gear to pinion gear backlash is being measured. If this measurement is found to be excessive, then the gears must be replaced, not the thrust washer.

Answer B is correct. Only technician B is correct. Backlash is measured to determine the wear between two gears. If this measurement is found to be excessive, then the gears will need to be replaced.

Answer C is incorrect. Only technician B is correct.

Answer D is incorrect. Technician B is correct.

16. Technician A says a worn U-joint may cause a squeaking noise. Technician B says a heavy vibration that only occurs during acceleration may be caused by a worn centering ball and socket on a double-Cardan U-joint. Who is right?

TASK D.1

    A. A only
    B. B only
    C. Both A and B
    D. Neither A nor B

    Answer A is incorrect. Technician B is also correct. The squeaking noise will increase as the vehicle speed increases.

    Answer B is incorrect. Technician A is also correct. The joint will not follow its intended arc, causing driveline imbalance.

    Answer C is correct. Both Technicians are correct.

    Answer D is incorrect. Both Technicians are correct.

17. Technician A says in order to measure differential carrier runout, the ring gear should be removed. Technician B says the side bearing caps must be installed. Who is correct?

TASK E.2.4

    A. A only
    B. B only
    C. Both A and B
    D. Neither A nor B

    Answer A is incorrect. Technician B is also correct.

    Answer B is incorrect. Technician A is also correct.

    Answer C is correct. Both Technicians are correct. The differential carrier runout is measured after the ring gear is removed. Place the carrier back in the rear end housing, and install side bearing caps and torque to specification. The maximum allowable carrier runout is 0.028 inches (0.07 mm).

    Answer D is incorrect. Both Technicians are correct.

18. The electric shift motor fails to operate on a transfer case. Technician A says the voltage should be checked at the shift motor electrical connector; if the voltage is equal to source voltage, then the shift motor needs replacing. Technician B says the power and ground sides of the shift motor should be tested for voltage drop before any component is replaced. Who is correct?

TASK F.11

    A. A only
    B. B only
    C. Both A and B
    D. Neither A nor B

    Answer A is incorrect. Just checking the voltage can lead to a misdiagnosis—just because a component has voltage getting to it does not mean there is enough amperage supplied to operate it. Twelve volts will travel through one single strand of a stranded copper wire.

    Answer B is correct. Only Technician B is correct. Doing a voltage-drop test is the best method for checking hidden resistance in a circuit. There should be B+ voltage right up to the motor and there should be 12 volts across the motor. A small drop across a connector or switch is allowed, usually limited to 0.1 to 0.2 volts. The circuit must be energized to do a voltage drop test.

    Answer C is incorrect. Only Technician B is correct.

    Answer D is incorrect. Technician B is correct.

TASK A.8

19. Technician A says that if clutch housing runout is out of limits, then shims can be used to correct the runout. Technician B says if shims are not available, they can be cut from shim stock. Who is correct?

A. A only

B. B. only

C. Both A and B

D. Neither A nor B

Answer A is incorrect. Technician B is also correct.

Answer B is incorrect. Technician A is also correct.

Answer C is correct. Both technicians are correct. If the clutch housing runout readings exceed the OEM specification, then the clutch housing can be shimmed or replaced. Shims are installed between the clutch housing and engine block at the location nearest the point with the lowest reading. If shims are not available, then they can be cut out of shim stock.

Answer D is incorrect. Both technicians are correct.

TASK D.2

20. A front-wheel drive vehicle has a clunking noise during acceleration or deceleration. Technician A says if the noise gets louder on a turn, then the problem is an inboard CV joint. Technician B says the problem is a worn-out inboard CV joint. Who is correct?

A. A only

B. B only

C. Both A and B

D. Neither A nor B

Answer A is incorrect. For a complaint of noise or vibration, the vehicle should be test driven. Accelerate with the wheels straight, then with the wheels turned left, then right. If the noise gets louder on a left or right turn, then the problem is probably a worn outboard CV joint.

Answer B is correct. Only technician B is correct. A vibration or a shudder indicates a sticking inboard CV joint. If the noise is a clucking noise, then this indicates a worn inboard CV joint.

Answer C is incorrect. Only technician B is correct.

Answer D is incorrect. Technician B is correct.

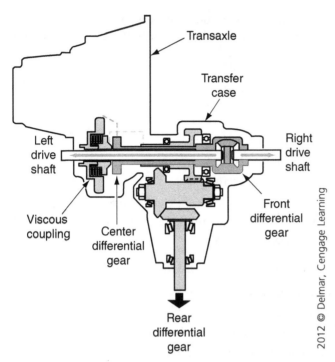

2012 © Delmar, Cengage Learning

21. In reference to the illustration, Technician A says if the center differential gear failed, then there would be no power flow to the rear wheels. Technician B says if the viscous coupling fails, then it must be replaced as a unit. Who is correct?

    A. A only

    B. B only

    C. Both A and B

    D. Neither A nor B

TASK F.11

Answer A is incorrect. Technician B is also correct.

Answer B is incorrect. Technician A is also correct.

Answer C is correct. Both technicians are correct. The viscous coupling allows the front and rear wheels to rotate at different speeds. If the viscous coupling fails, then it must be replaced as a unit. The center differential gear transfers power to the rear wheels. If this gear fails, then no power will be transmitted to the rear wheels.

Answer D is incorrect. Both technicians are correct.

22. A transmission has a growling noise in neutral, with the clutch engaged. Technician A says the output shaft bearing could be the problem. Technician B says the input shaft bearing could be the problem. Who is correct?

    A. A only

    B. B only

    C. Both A and B

    D. Neither A nor B

TASK B7

Answer A is incorrect. The output shaft is not rotating in neutral, so the output shaft bearing cannot be making the noise.

Answer B is correct. Only technician B is correct. The input shaft is rotating any time the engine is running and the clutch is engaged.

Answer C is incorrect. Only technician B is correct.

Answer D is incorrect. Technician B is correct.

TASK C.4

23. Technician A says a misaligned engine and transaxle cradle may cause driveline vibrations. Technician B says a misaligned engine and transaxle cradle may affect front suspension angles. Who is right?

    A.   A only
    B.   B only
    C.   Both A and B
    D.   Neither A nor B

    Answer A is incorrect. Technician B is also correct.

    Answer B is incorrect. Technician A is also correct.

    Answer C is correct. Both technicians are correct. If the cradle is removed, then it must be realigned properly, or else both driveline angle and front-end alignment angles will be affected.

    Answer D is incorrect. Both technicians are correct.

TASK D.5

24. To measure drive shaft runout, you should do all of the following EXCEPT:

    A.   Place the vehicle transmission in neutral.
    B.   Use a magnetic-base dial indicator.
    C.   Roll the drive shaft on a flat surface to check for damage.
    D.   Clean the drive shaft surface for an accurate runout check.

    Answer A is incorrect. The transmission should be placed in neutral to measure runout so that the technician can rotate the drive shaft.

    Answer B is incorrect. A magnetic-base dial indicator is typically used. While a clamp-type indicator may be needed on a select few, most use a magnetic-base dial indicator.

    Answer C is correct. Rolling the drive shaft on a flat surface is not an acceptable way of checking the drive shaft.

    Answer D is incorrect. The drive shaft must be cleaned with a wire brush on the area where the dial indicator stylus will ride.

TASK E.3.2

25. Technician A says the proper level of lubricant on a limited-slip differential is one finger joint in the fill hole. Technician B says some limited-slip differentials require a special oil additive. Who is correct?

    A.   A only
    B.   B only
    C.   Both A and B
    D.   Neither A nor B

    Answer A is incorrect. Most manufacturers specify that the proper lubrication level is even with the bottom of the fill hole.

    Answer B is correct. Only technician B is correct. Some manufacturers require the addition of a special additive to the lubrication. This additive is a friction modifier that allows the limited-slip differential fiber disc to slip during cornering maneuvers.

    Answer C is incorrect. Only technician B is correct.

    Answer D is incorrect. Technician B is correct.

26. Excessive noise coming from the transfer case may be caused by all of the following EXCEPT:

    A. Low fluid level
    B. Misalignment of the transfer-case drive chain
    C. A worn universal joint
    D. A damaged output-shaft bearing

TASK F.1

Answer A is incorrect. A low fluid level would allow bearings to overheat and cause growling.

Answer B is incorrect. If the transfer case drive chain is the location of the misalignment, then it would make noise.

Answer C is correct. A worn universal joint would make noise, but not the transfer case.

Answer D is incorrect. A damaged output-shaft bearing would make noise any time the transfer case is rotating.

27. A transaxle has brass flakes in its oil during an oil change. Technician A says this is a sign that the bearings are beginning to chip and flake. Technician B says the synchronizers are beginning to flake and wear. Who is correct?

    A. A only
    B. B only
    C. Both A and B
    D. Neither A nor B

TASK C.1

Answer A is incorrect. If the bearing was flaking, then the oil would show flakes of steel material, not brass.

Answer B is correct. Only technician B is correct. The blocker ring in most transaxles is made of brass. As the blocker ring wears, the transaxle oil may show signs of brass particles and flakes.

Answer C is incorrect. Only technician B is correct.

Answer D is incorrect. Technician B is correct.

28. Technician A says that if a clutch noise only happens when the pedal is all of the way to the floor, then the release bearing is probably the cause. Technician B says the noise that happens only when the engine is idling in neutral with the clutch engaged is probably due to an input-shaft bearing. Who is correct?

    A. A only
    B. B only
    C. Both A and B
    D. Neither A nor B

TASK A.1

Answer A is incorrect. If the release bearing is worn and growling, then the noise would begin as soon as the free travel is taken up in the pedal. The noise would start at the beginning of clutch disengagement. If clutch noise is heard only when the clutch pedal is all of the way to the floor, then the problem is probably the pilot bearing.

Answer B is correct. Only technician B is correct. The input-shaft bearing is rotating any time the clutch is engaged and the engine is running.

Answer C is incorrect. Only technician B is correct.

Answer D is incorrect. Technician B is correct.

TASK E.4.1

29. A vehicle has a growling noise from the rear end and gets louder on a sharp right turn. Technician A says the vehicle has an axle bearing beginning to fail. Technician B says that the differential side bearing is causing the noise. Who is correct?

    A.   A only
    B.   B only
    C.   Both A and B
    D.   Neither A nor B

    Answer A is correct. Only technician A is correct. A bad axle bearing will get louder on a sharp turn because this loads the axle bearing even more.

    Answer B is incorrect. A bad carrier bearing makes a low constant growl that will not change when turning.

    Answer C is incorrect. Only technician A is correct.

    Answer D is incorrect. Technician A is correct.

TASK A.3

30. Technician A says the hydraulic clutch system can be bled by putting fluid into the slave cylinder bleed screw. Technician B says that a hydraulic clutch system can be bled by letting fluid out of the slave cylinder bleed screw. Who is correct?

    A.   A only
    B.   B only
    C.   Both A and B
    D.   Neither A nor B

    Answer A is incorrect. Technician B is also correct.

    Answer B is incorrect. Technician A is also correct.

    Answer C is correct. Both technicians are correct. There are several ways to bleed the clutch hydraulic system. One method is to fill the reservoir with fluid and open the bleeder screw on the slave cylinder and let it gravity-bleed. Another method would be to pressure-bleed the system through the bleeder screw.

    Answer D is incorrect. Both technicians are correct.

TASK E.1.2

31. A vehicle has a vibration above 30 mph in the rear. Technician A says the driveline and companion flanges should be checked for excessive runout. Technician B advises to disconnect the flanges and reconnect them 180 degrees apart if runout is found. Who is correct?

    A.   A only
    B.   B only
    C.   Both A and B
    D.   Neither A nor B

    Answer A is incorrect. Technician B is also correct.

    Answer B is incorrect. Technician A is also correct.

    Answer C is correct. Both technicians are correct. A driveline vibration is more noticeable at speeds above 30 mph. If a vibration is felt above 30 mph, then the runout of the driveline and companion flanges should be checked. If runout is found, then the flanges should be disconnected from each other. Each one should be rotated 180 degrees apart, and then runout should be checked again.

    Answer D is incorrect. Both technicians are correct.

32. Technician A says the cluster gears are a major component of the transmission lubrication system. Technician B says some transmissions use oiling troughs to guide the oil to critical areas. Who is correct?

    A. A only
    B. B only
    C. Both A and B
    D. Neither A nor B

    TASK B.16

    Answer A is incorrect. Technician B is also correct.

    Answer B is incorrect. Technician A is also correct.

    Answer C is correct. Both technicians are correct. The transmission cluster gears run in a bath of lubrication; as they rotate, they pick up the oil and throw it around in the case and into oiling troughs that guide the oil to critical areas. The oil cools, lubricates, and cleans the transmission.

    Answer D is incorrect. Both technicians are correct.

33. All of the following are part of the rear oil-seal replacement procedure on a transmission EXCEPT:

    A. Remove the drive shaft.
    B. Lubricate the lip of the new seal.
    C. Drive a new seal in place with a hammer and brass punch.
    D. Torque universal joint cap screws.

    TASK B.3

    Answer A is incorrect. The drive shaft must be removed in order to gain access to the transmission rear seal.

    Answer B is incorrect. The lip of all seals should be lubricated with whatever type of oil is used in the transmission when installing.

    Answer C is correct. A seal driver should be used to install new seals to prevent damage to the seal housing.

    Answer D is incorrect. The last step in output seal replacement is to reinstall the drive shaft and torque the universal joint cap screws to specification.

34. The transfer clutch in an all-wheel drive vehicle takes the place of which one of the following?

    A. Reduction gears
    B. Torque converter
    C. Interaxle differential
    D. Drive chain

    TASK F.11

    Answer A is incorrect. Reduction gears are used in four-wheel drive transfer cases for low range.

    Answer B is incorrect. A torque converter is used in an automatic transmission.

    Answer C is correct. An all-wheel drive vehicle can use an interaxle differential, viscous coupler, or transfer clutch.

    Answer D is incorrect. The drive chain is used in a transfer case to connect the front axle to the rest of the drive train.

TASK E.2.1

35. A rear-wheel drive vehicle is being diagnosed for axle noise. Technician A says grinding noise only heard when going around a curve or turning is probably in one of the axle bearings. Technician B says a clicking noise only heard when going around a curve or turning is probably in the outer constant-velocity joint. Who is correct?

A. A only

B. B only

C. Both A and B

D. Neither A nor B

Answer A is incorrect. An axle bearing will have a growling noise that gets louder on turns.

Answer B is incorrect. A rear-wheel drive vehicle does not use constant-velocity joints in the front.

Answer C is incorrect. Neither technician is correct.

Answer D is correct. Neither technician is correct. A noise that is only heard when cornering or going around a curve can be narrowed down to the differential. It can be the pinion gears, side gears, or both.

TASK B.15

36. When checking a diagnostic trouble code (DTC) for the output shaft speed sensor, which of the following should NOT be done after reading the code?

A. Install a new sensor.

B. Inspect the harness for continuity.

C. Check the service manual.

D. Inspect the connector.

Answer A is correct. The sensor should not be replaced until it is confirmed to be defective.

Answer B is incorrect. The harness should always be inspected when troubleshooting speed sensor system fault.

Answer C is incorrect. The service manual is a good place to start your diagnostics. With most DTCs, there will be a diagnostic flow chart in the service manual to assist in the diagnosis.

Answer D is incorrect. A lot of diagnostic trouble codes can be traced to the connector. Always perform a thorough visual inspection of the sensor connector.

TASK C.1

37. A 4-speed transaxle makes a clunking noise when driven in first gear and reverse gear. Technician A says that the gear on the countershaft could be the problem. Technician B says that the reverse idler gear could be the problem. Who is right?

A. A only

B. B only

C. Both A and B

D. Neither A nor B

Answer A is correct. Only technician A is correct. The power flow for first and reverse gears both go through the countershaft gears.

Answer B is incorrect. The only time the power flows through the reverse idler is when reverse gear is engaged. If the clunk is in first and reverse gears, it cannot be the reverse idler.

Answer C is incorrect. Only technician A is correct.

Answer D is incorrect. Technician A is correct.

38. An electronic transfer case does not engage. Which of the following would be the LEAST LIKELY cause?

    A. A bad electric shift motor

    B. A blown fuse

    C. The 4WD engage switch

    D. A rusted linkage

TASK F.11

    Answer A is incorrect. If the electric shift motor malfunctions, then this could prevent transfer case engagement.

    Answer B is incorrect. The fuse protects electric circuits from current overload. A blown fuse would prevent transfer case engagement. If a blown fuse is found, then the entire circuit should be checked to determine the cause.

    Answer C is incorrect. The four-wheel drive switch is the input to the computer requesting transfer case engagement. If the switch fails, then the transfer case would not engage.

    Answer D is correct. An electronic transfer case does not use mechanical linkage.

39. The best way to check mating surfaces for warpage on a transaxle case is to use which of the following?

    A. Straight edge

    B. Dial indicator

    C. Micrometer

    D. Flat surface

TASK C.3

    Answer A is correct. Using a straight edge and a feeler gauge, the technician can check for any low spots in the transaxle case. Surface warpage of 0.002 in (0.5mm) or smaller is acceptable.

    Answer B is incorrect. A dial indicator is used for end-play, runout, and sometimes selective shim calculations.

    Answer C is incorrect. A micrometer is used to measure the thickness of a component.

    Answer D is incorrect. The flat surface is what the technician is measuring.

40. Technician A says that when a wheel bearing is replaced, the bearing race should also be replaced. Technician B says if the wheel bearing is a unitized bearing, then the race does not need replacing. Who is right?

    A. A only

    B. B only

    C. Both A and B

    D. Neither A nor B

TASK F.9

    Answer A is incorrect. Technician B is also correct.

    Answer B is incorrect. Technician A is also correct.

    Answer C is correct. Both technicians are correct. Any time a wheel bearing is replaced, the bearing race should also be replaced. Using the old bearing race with a new bearing can cause the new bearing to fail prematurely. If the wheel bearing is a unitized bearing, then it is sealed and only serviced as a complete assembly.

    Answer D is incorrect. Both technicians are correct.

PREPARATION EXAM 4—ANSWER KEY

1.	B	21.	A
2.	A	22.	C
3.	A	23.	A
4.	A	24.	D
5.	D	25.	B
6.	A	26.	A
7.	B	27.	C
8.	C	28.	B
9.	B	29.	D
10.	D	30.	D
11.	B	31.	B
12.	C	32.	C
13.	C	33.	A
14.	C	34.	D
15.	C	35.	B
16.	C	36.	D
17.	D	37.	C
18.	B	38.	C
19.	A	39.	A
20.	C	40.	C

PREPARATION EXAM 4—EXPLANATIONS

TASK A.1

1. Technician A says to put the transmission in low gear to test for a slipping clutch. Technician B says that as a mechanical linkage clutch disc wears, the free travel becomes less. Who is correct?

   A. A only
   B. B only
   C. Both A and B
   D. Neither A nor B

   Answer A is incorrect. To test for a slipping clutch, put the transmission in high gear, then engage the clutch. If the vehicle stalls, then the clutch is not slipping.

   Answer B is correct. Only technician B is correct. As the clutch disc wears, the pressure plate gets closer to the flywheel, deceasing free travel.

   Answer C is incorrect. Only technician B is correct.

   Answer D is incorrect. Technician B is correct.

2. Technician A says that repeated extension-housing seal failures may be caused by a worn extension-housing bushing. Technician B says that a worn countershaft bearing may cause an extension-housing bushing to wear prematurely. Who is correct?

TASK B.3

    A. A only
    B. B only
    C. Both A and B
    D. Neither A nor B

    Answer A is correct. Only technician A is correct. A worn extension-housing bushing allows the drive shaft yoke to have excessive runout; this will damage the extension housing seal and will allow it to leak.

    Answer B is incorrect. The countershaft is not in direct contact with the extension housing bushing or seal, and it will have no effect on the components.

    Answer C is incorrect. Only technician A is correct.

    Answer D is incorrect. Technician A is correct.

3. A transaxle has had repeated drive-axle seal leakage and replacements. Technician A says this problem may be caused by a plugged transaxle vent. Technician B says this problem may be caused by a worn outer constant-velocity joint. Who is correct?

TASK C.3

    A. A only
    B. B only
    C. Both A and B
    D. Neither A nor B

    Answer A is correct. Only technician A is correct. If the vent gets plugged, then the transaxle cannot vent the excess pressure that it builds up as the lubricating oil heats up. This excess pressure can push the oil past a good seal.

    Answer B is incorrect. The outer constant-velocity joint does not come in contact with the drive axle seal. The inner CV joint is the one in contact.

    Answer C is incorrect. Only technician A is correct.

    Answer D is incorrect. Technician A is correct.

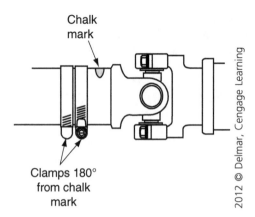

Chalk mark

Clamps 180° from chalk mark

2012 © Delmar, Cengage Learning

TASK D.4

4. Technician A says the hose clamps in the illustration are for drive shaft balance. Technician B says the chalk mark shows the runout location. Who is correct?

   A. A only

   B. B only

   C. Both A and B

   D. Neither A nor B

   Answer A is correct. Only technician A is correct. The hose clamps are used for drive shaft balance.

   Answer B is incorrect. The chalk mark is used to reference the heavy spot during the balancing of the drive shaft. It is not used for runout.

   Answer C is incorrect. Only technician A is correct.

   Answer D is incorrect. Technician A is correct.

TASK E.3.3

5. All of the following statements about a limited-slip differential are true EXCEPT:

   A. Fiber plates are splined to the side gear.

   B. A special lubricant additive is needed.

   C. Each clutch pack uses a preload spring.

   D. The breakaway torque is checked with an inch-pound torque wrench.

   Answer A is incorrect. The fiber plates are splined to the side gears, alternating steel and fiber.

   Answer B is incorrect. A special friction modifier is used in limited-slip differentials with clutch discs.

   Answer C is incorrect. The preload spring keeps the needed pressure on the clutch packs to allow the limited slip.

   Answer D is correct. The breakaway torque of a limited-slip differential is checked using a foot pound torque wrench, not an inch pound.

6. All of the following statements regarding 4WD front-drive axles and joints are true EXCEPT:

TASK F.7

A. The joint should be coated with good marine-grade NLG 2 grease.

B. A worn CV joint may cause a clicking noise when cornering.

C. The inner tripod joint is held on the axle shaft with a cir-clip.

D. A special tool is required to install the boot clamp.

Answer A is correct. A CV joint requires a special grease that comes premeasured. All of the grease supplied should be used.

Answer B is incorrect. A clicking noise from the front drive axle when cornering is usually the result of a bad outer CV joint.

Answer C is incorrect. The inner CV joint is held in place with either a cir-clip or a snap ring. Some inner CV joints bolt directly to the flange.

Answer D is incorrect. A special band is used to hold the CV boot in place. A special tool is required to clamp the boot to the needed amount to prevent clamp slippage.

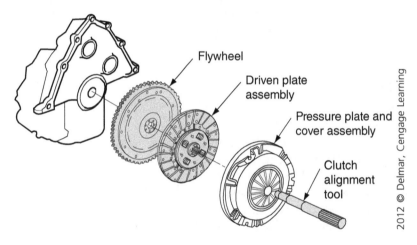

7. All of the following are true about the clutch alignment tool EXCEPT:

TASK A.5

A. Align the pilot bearing with the clutch hub opening.

B. Align the pressure plate with the flywheel.

C. Hold the clutch disc in alignment during pressure plate installation.

D. You can make one out of an old transmission input shaft.

Answer A is incorrect. The clutch alignment tool is used to align the pilot bearing with the clutch hub opening for easy transmission installation.

Answer B is correct. Guide studs are used to help align the pressure plate with the flywheel during reassembly.

Answer C is incorrect. The clutch alignment tool holds the clutch disc in alignment with the pilot bearing during reassembly.

Answer D is incorrect. Many technicians make their own clutch alignment tool out of an old transmission input shaft.

TASK B.15

8. Technician A says that a back-up lamp switch is a normally open switch. Technician B says that a back-up lamp switch can have power to the switch in run and is closed when the vehicle is shifted into reverse gear. Who is correct?

    A.   A only

    B.   B only

    C.   Both A and B

    D.   Neither A nor B

Answer A is incorrect. Technician B is also correct.

Answer B is incorrect. Technician A is also correct.

Answer C is correct. Both technicians are correct. The back-up light switch is a normally open switch and is closed when the vehicle is put in reverse gear. The power that is supplied to the back-up light bulbs is supplied from the ignition switch in the run position, flows to the back-up light switch, and then to the back-up lights.

Answer D is incorrect. Both technicians are correct.

TASK C.12

9. Technician A says vehicle speed sensors can be tested with an ammeter. Technician B says when you check a vehicle speed sensor signal, you should use an oscilloscope while the vehicle is being driven or run on the rack. Who is correct?

    A.   A only

    B.   B only

    C.   Both A and B

    D.   Neither A nor B

Answer A is incorrect. A VSS can be a Hall-effect sensor or a magnetic pulse generator. Neither type can be tested with an ammeter. A magnetic pulse generating sensor can be checked for resistance; its AC voltage can also be checked with a voltmeter or a lab scope can be used to view the pattern. A Hall-effect sensor cannot be checked for resistance because it puts out a pulsating DC signal. Using a lab scope to view the pattern is the preferred method.

Answer B is correct. Only technician B is correct. One of the most effective methods for diagnosing a VSS is by hooking up a lab scope and driving the vehicle or just running the vehicle on a rack in drive.

Answer C is incorrect. Only technician B is correct.

Answer D is incorrect. Technician B is correct.

TASK D.5

10. Technician A says the dial indicator should be positioned near the rear of the drive shaft to measure drive shaft runout. Technician B says if the drive shaft runout is out of limits, then the drive shaft may be straightened using a hydraulic press and straight edge. Who is correct?

    A.   A only

    B.   B only

    C.   Both A and B

    D.   Neither A nor B

Answer A is incorrect. The drive shaft should be measured for runout in the center and three inches from each end.

Answer B is incorrect. Technicians do not straighten drive shafts using a hydraulic press and straight edge.

Answer C is incorrect. Neither technician is correct.

Answer D is correct. Neither technician is correct. When measuring drive shaft runout, the drive shaft should be cleaned with a wire brush at the center and three inches from each end. Then place the dial indicator at each of the areas cleaned, and rotate one complete revolution for each area being measured. If the drive shaft is found out of limits, then the drive shaft must be replaced.

11. Technician A says that excessive pinion-nut torque may cause a clunking noise during acceleration or deceleration. Technician B says that excessive pinion-nut torque may cause a growling noise with the vehicle in motion. Who is correct?

TASK E.1.1

   A.   A only
   B.   B only
   C.   Both A and B
   D.   Neither A nor B

   Answer A is incorrect. An undertightened pinion nut would cause a clunking noise on acceleration and deceleration, not an overtightened nut.

   Answer B is correct. Only technician B is correct. An overtightened pinion nut would cause excessive preload on the bearings, which would lead to the bearings growling.

   Answer C is incorrect. Only technician B is correct.

   Answer D is incorrect. Technician B is correct.

12. When a torque management transfer case is operating in automatic four-wheel drive (A4WD) and the front drive shaft begins to turn faster than the rear drive shaft, what happens?

   A.   The computer energizes the viscous coupling.
   B.   The computer reduces the throttle opening.
   C.   The computer increases the clutch coil duty-cycle.
   D.   The computer reduces the torque to the rear wheels.

TASK F.11

   Answer A is incorrect. A viscous coupling is a sealed clutch filled with a viscous fluid that transfers power as the fluid is heated but is not computer-controlled.

   Answer B is incorrect. The computer does not reduce throttle opening. Some traction control systems reduce power to the engine when spinning is detected but do not reduce throttle opening.

   Answer C is correct. In a torque-management transfer case, the computer will increase the clutch coil duty cycle, which will increase clutch-pack holding power to control the spinning front drive shaft.

   Answer D is incorrect. Reducing the torque to the rear wheels will cause the front to spin even more.

13. Technician A says if too much material is machined away during flywheel resurfacing, then the torsion springs on the clutch disc may contact the flywheel bolts, resulting in noise while engaging and disengaging. Technician B says if excessive material is removed when the flywheel is resurfaced, then the slave cylinder rod may not have enough travel to release the clutch properly. Who is correct?

TASK A.7

   A.   A only
   B.   B only
   C.   Both A and B
   D.   Neither A nor B

   Answer A is incorrect. Technician B is also correct.

   Answer B is incorrect. Technician A is also correct.

   Answer C is correct. Both technicians are correct. Each time the flywheel is machined, more material is removed. This causes the clutch disc and pressure plate to be closer to the engine. If enough material is removed, then the flywheel mounting bolts can contact the clutch disc. While the clutch disc and pressure plate move closer to the engine each time the flywheel is machined, the distance needed by the clutch slave cylinder increases. The hydraulics can compensate for this in a limited distance, but if excessive material is removed, then it will not be able to compensate, preventing proper clutch disengagement.

   Answer D is incorrect. Both technicians are correct.

TASK B.2

14. The shift lever adjustment on a transmission with external linkage is usually performed with the transmission in which gear?

    A. First

    B. Second

    C. Neutral

    D. Reverse

    Answer A is incorrect. The transmission linkage is adjusted with the transmission in neutral; a lock or pin is usually used to hold the shift lever in the correct position.

    Answer B is incorrect. The transmission linkage is adjusted with the transmission in neutral; a lock or pin is usually used to hold the shift lever in the correct position.

    Answer C is correct. The transmission linkage is adjusted with the transmission in neutral; a lock or pin is usually used to hold the shift lever in the correct position.

    Answer D is incorrect. The transmission linkage is adjusted with the transmission in neutral; a lock or pin is usually used to hold the shift lever in the correct position.

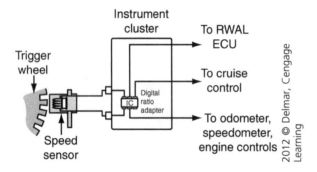

TASK C.12

15. Technician A says if the speedometer shown does not work on a vehicle with a transaxle, then check the signal at the vehicle speed sensor. Technician B says the trigger wheel could be defective. Who is correct?

    A. A only

    B. B only

    C. Both A and B

    D. Neither A nor B

    Answer A is incorrect. Technician B is also correct.

    Answer B is incorrect. Technician A is also correct.

    Answer C is correct. Both technicians are correct. On a vehicle with a transaxle, the speedometer can be mechanically or electrically controlled. On electrically controlled speedometers, the speed input comes from the vehicle speed sensor. This sensor can be a Hall-effect sensor or a magnetic pulse-generating sensor. Either type requires a trigger wheel rotating to produce a signal. If the speedometer is not working, then one of the first checks would be for a signal at the sensor itself.

    Answer D is incorrect. Both technicians are correct.

16. A rear-wheel drive vehicle has a vibration that increases in relation to vehicle speed. Technician A says the balance weight may have fallen off the drive shaft. Technician B says it might have a wheel out of balance. Who is correct?

TASK D.4

    A. A only
    B. B only
    C. Both A and B
    D. Neither A nor B

    Answer A is incorrect. Technician B is also correct.

    Answer B is incorrect. Technician A is also correct.

    Answer C is correct. Both technicians are correct. When a drive shaft is manufactured, one of the last steps is to balance the shaft. Weights are welded to the shaft to balance it. Road debris can sometimes knock these weights off, causing an imbalance condition. An imbalanced drive shaft will cause a vibration that increases with vehicle speed. A tire or a wheel out of balance will cause the same effect.

    Answer D is incorrect. Both technicians are correct.

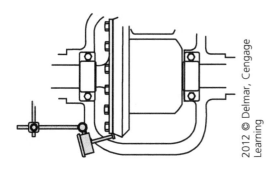

2012 © Delmar, Cengage Learning

17. Excessive ring-gear runout on the dial indicator shown in the illustration may be caused by excessive:

TASK E.1.3

    A. Pinion-bearing preload
    B. Side gear end-play
    C. Side-bearing preload
    D. Differential case/carrier runout

    Answer A is incorrect. Excessive pinion-bearing preload would cause the pinion bearing to make a growling noise and fail.

    Answer B is incorrect. Problems with the side gear end-play would cause noises on turns, but not excessive ring-gear runout.

    Answer C is incorrect. Excessive side-bearing preload would cause the side bearings to produce a low-pitch growling noise when driving above 10–15 mph.

    Answer D is correct. Since the ring gear bolts to the differential case, excessive runout at the case would affect the runout at the ring gear.

TASK F.1

18. Excessive noise coming from the transfer case may be caused by all of the following EXCEPT:

   A. Low transfer-case fluid level
   B. A worn transfer-case universal joint
   C. Misalignment of the transfer-case drive chain
   D. A damaged transfer-case output shaft bearing

   Answer A is incorrect. Running the transfer case low on fluid can overheat gears and bearings, causing excessive noise.

   Answer B is correct. The universal joints used on a four-wheel drive vehicle are external and not inside the transfer case.

   Answer C is incorrect. If the drive chain were to get out of alignment, then it could cause noise when in four-wheel drive.

   Answer D is incorrect. A damaged output shaft bearing would make noise any time the vehicle was moving.

TASK A.4

19. Technician A says the clutch pedal freeplay adjustment sets the distance between the release bearing and the pressure-plate fingers. Technician B says a worn release bearing makes the most noise when the clutch is engaged. Who is correct?

   A. A only
   B. B only
   C. Both A and B
   D. Neither A nor B

   Answer A is correct. Only technician A is correct. The clutch pedal freeplay adjustment ensures there is sufficient clearance between the release bearing and the pressure-plate fingers. As the clutch disc wears, this distance decreases.

   Answer B is incorrect. A worn release bearing would make the most noise as the clutch is disengaged or the clutch pedal is engaged.

   Answer C is incorrect. Only technician A is correct.

   Answer D is incorrect. Technician A is correct.

TASK B.12

20. Bearing preload in a transmission is

   A. To test a bearing before it is installed
   B. The amount of pressure a bearing can handle while under a load
   C. The amount of pressure applied to a bearing upon assembly of the transmission
   D. Not adjustable

   Answer A is incorrect. A bearing is tested by feel and visual inspection

   Answer B is incorrect. The amount of pressure a bearing can handle is determined by the bearing size and type.

   Answer C is correct. When a bearing is preloaded, a specific amount of pressure is applied to the bearing using shims at the time of transmission assembly.

   Answer D is incorrect. Bearing preload is adjustable using selective shims.

21. A four-speed manual transaxle jumps out of third gear. Technician A says the shift-rail detent spring tension on the 3-4 shift rail may be weak. Technician B says there may be excessive wear on the fourth-speed gear clutching teeth. Who is correct?

    A. A only

    B. B only

    C. Both A and B

    D. Neither A nor B

    TASK C.6

    Answer A is correct. Only technician A is correct. The purpose of the detent spring and detent is to hold the transaxle in the selected gear. A weak or broken 3-4 detent spring would allow the transaxle to jump out of gear at times.

    Answer B is incorrect. Excessive wear on the clutching teeth of a speed gear can allow a transaxle to jump out of gear, but it would only affect the gear with the wear. In this case, it would allow fourth gear to jump out, not third.

    Answer C is incorrect. Only technician A is correct.

    Answer D is incorrect. Technician A is correct.

22. Technician A says the front wheel bearings on a vehicle with four-wheel drive and tapered front wheel bearings should be adjusted with 0.001–0.005 of end-play. Technician B says some late-model vehicles use unitized wheel bearings that are not adjustable. Who is correct?

    A. A only

    B. B only

    C. Both A and B

    D. Neither A nor B

    TASK D.7

    Answer A is incorrect. Technician B is also correct.

    Answer B is incorrect. Technician A is also correct.

    Answer C is correct. Both technicians are correct. A vehicle with tapered front wheel bearings needs end-play on the bearings to prevent the bearings from overheating. Excessive end-play would be just as bad. Most manufacturers specify an end-play of 0.001–0.005 inches (0.025–0.127 mm).

    Answer D is incorrect. Both technicians are correct.

23. An axle shaft is removed from the differential. Next it should be:

    A. Placed standing straight up on the floor

    B. Checked for axle flange runout after removal

    C. Left under the vehicle

    D. Placed flat in a horizontal position on the floor

    TASK E.4.3

    Answer A is correct. When the axle is removed from the differential, it should be placed in an upright position.

    Answer B is incorrect. If axle flange runout is checked, then it should be checked before removal.

    Answer C is incorrect. The axle should not be left under the vehicle. This would prevent it from being stood up in the vertical position.

    Answer D is incorrect. The axle shaft should not be placed flat on the floor in a horizontal position.

TASK F.1

24. Technician A says the transfer case can be shifted into 4L at any speed in an electronically shifted transfer case. Technician B says the transfer case can be shifted into 4L with the transmission in any gear. Who is correct?

    A.   A only
    B.   B only
    C.   Both A and B
    D.   Neither A nor B

    Answer A is incorrect. The vehicle speed must be below 3 mph before the transfer case will shift into 4L.

    Answer B is incorrect. The transmission must be in neutral before the transfer case will shift into 4L.

    Answer C is incorrect. Neither technician is correct.

    Answer D is correct. Neither technician is correct. The transfer case can only be shifted in low range when the vehicle speed is below 3 mph and in neutral, regardless of whether it is electronic or mechanical. Forcing a shift into low range above this speed while in gear can cause damage to the transfer case.

TASK F.7

25. Technician A says that a worn ring gear on the flywheel can cause clutch chatter. Technician B says that a worn ring gear on the flywheel can affect the starter engagement. Who is correct?

    A.   A only
    B.   B only
    C.   Both A and B
    D.   Neither A nor B

    Answer A is incorrect. The ring gear is for the starter to engage to rotate the engine for startup.

    Answer B is correct. Only technician B is correct. The ring gear should always be inspected when the clutch is being replaced. Worn ring-gear teeth can cause a grinding or raking noise on start up

    Answer C is incorrect. Only technician B is correct.

    Answer D is incorrect. Technician B is correct.

TASK B.13

26. Which of the following components is LEAST LIKELY to be replaced when replacing the extension housing?

    A.   The countergear shaft
    B.   The tail-housing gasket
    C.   The speedometer O-ring
    D.   The output shaft seal

    Answer A is correct. Only technician A is correct. The countershaft is located internally in the transmission. The transmission would need to be disassembled to replace the countershaft.

    Answer B is incorrect. A new gasket should always be used when replacing the extension housing.

    Answer C is incorrect. The speedometer drive is located on the output shaft. The speedometer drive would be removed when replacing the extension housing, so a new O-ring would be required.

    Answer D is incorrect. A new seal is always required when removing the extension housing.

27. Technician A says that if the intermediate shaft gear is worn, then the mating gear on the input shaft should be replaced with the new intermediate shaft. Technician B says that if the intermediate shaft gear is worn, then the case and all of the related bearings should be checked. Who is correct?

TASK C.7

   A. A only
   B. B only
   C. Both A and B
   D. Neither A nor B

   Answer A is incorrect. Technician B is also correct.

   Answer B is incorrect. Technician A is also correct.

   Answer C is correct. Both technicians are correct. Any time the intermediate shaft or countershaft gear is worn, the mating gear on the input shaft should be replaced. The case and all related bearings should also be checked at this time.

   Answer D is incorrect. Both technicians are correct.

28. Technician A says the drive-shaft carrier bearing can cause noises in neutral if the vehicle is stopped and clutch is disengaged. Technician B says that some center-support bearings need to be lubricated. Who is correct?

TASK D.3

   A. A only
   B. B only
   C. Both A and B
   D. Neither A nor B

   Answer A is incorrect. The only time the carrier (center-support) bearing is rotating is when the vehicle is in motion. The carrier bearing cannot make noise with the vehicle sitting still.

   Answer B is correct. Only technician B is correct. Some carrier (center-support) bearings are sealed and cannot be lubricated, while others have a grease zerk and can be lubricated.

   Answer C is incorrect. Only technician B is correct.

   Answer D is incorrect. Technician B is correct.

29. Technician A says that preload on the pinion gear should be measured with a dial indicator. Technician B says that pinion gear preload can also be measured with a flat feeler gauge. Who is correct?

TASK E.1.6

   A. A only
   B. B only
   C. Both A and B
   D. Neither A nor B

   Answer A is incorrect. Dial indicators are used to measure end-play, backlash, and runout, not preload.

   Answer B is incorrect. A feeler gauge is used to measure distance like end-play and clearances, not preload.

   Answer C is incorrect. Neither technician is correct.

   Answer D is correct. Neither technician is correct. To measure pinion bearing preload, an inch pound dial-type torque wrench is used to measure the rotating torque.

TASK F.12

30. All of the following are part of a tire inspection on a four-wheel drive vehicle EXCEPT:

A. Aspect ratio

B. Nominal width

C. Load rating

D. Brand of tire

Answer A is incorrect. The aspect ratio of a tire is the tire height compared to width. This ratio is important because different aspect-ratio tires have different heights.

Answer B is incorrect. Nominal width is important because different-width tires on a four-wheel drive vehicle will have a different aspect ratio, which affects tire diameter.

Answer C is incorrect. The load rating is part of a tire inspection, due to safety concerns.

Answer D is correct. The brand of tire is not a concern during a tire inspection, as long as the speed, load, and size are correct for the vehicle.

TASK A.4

31. A vehicle has a squealing noise coming from the clutch area when the clutch is disengaged. Technician A says the release bearing is probably dry and needs to be lubricated. Technician B says the damage has been done and the release bearing should be replaced. Who is correct?

A. A only

B. B only

C. Both A and B

D. Neither A nor B

Answer A is incorrect. Most release bearings do not have a grease fitting that allows for lubrication.

Answer B is correct. Only technician B is correct. A noisy release bearing needs to be replaced once it starts making noise. Most release bearings do not have a grease fitting that allow for lubrication.

Answer C is incorrect. Only technician B is correct.

Answer D is incorrect. Technician B is correct.

TASK B.5

32. A transmission is being disassembled for overhaul. Technician A says care should be taken to thoroughly inspect each component for damage as it is disassembled. Technician B says some transmission components have index marks that must be aligned when reassembled. Who is correct?

A. A only

B. B only

C. Both A and B

D. Neither A nor B

Answer A is incorrect. Technician B is also correct.

Answer B is incorrect. Technician A is also correct.

Answer C is correct. Both technicians are correct. During a transmission rebuild, all components should be thoroughly inspected as the transmission is disassembled. Pay close attention to bolt lengths as well as any index marks that some components might have.

Answer D is incorrect. Both technicians are correct.

33. Technician A says that a bent shift rail should be replaced with a new shift rail. Technician B says that the shift rail may be heated and bent back into its original position using a dial indicator. Who is correct?

    TASK C.6

    A. A only
    B. B only
    C. Both A and B
    D. Neither A nor B

    Answer A is correct. Only technician A is correct. A shift rail must be replaced when found to be bent. Shift rails must fit precisely and cannot be repaired.

    Answer B is incorrect. There is no repair procedure for the shift rail except replacement.

    Answer C is incorrect. Only technician A is correct.

    Answer D is incorrect. Technician A is correct

34. All of the following are true about removing a differential assembly EXCEPT:

    TASK E.2.2

    A. The axle shafts must be removed.
    B. The bearing caps should be marked to the housing.
    C. The bearing races and shim packs should not be mixed up.
    D. The pinion gear always stays in the axle housing.

    Answer A is incorrect. The axle shafts must be removed in order to remove the differential.

    Answer B is incorrect. The bearing caps should always be marked before removing to ensure they are replaced in their original position.

    Answer C is incorrect. When removing and disassembling a differential, keep all races and shims marked and separated to ensure correct reinstallation.

    Answer D is correct. Some differentials are removed as a complete unit, while others are an integral part of the axle housing.

35. A remote vent on a transfer case is used to do what?

    TASK F.10

    A. Add lubricating oil to the transfer case
    B. Prevent water from entering the transfer case
    C. Keep lubricant in the transfer case
    D. Increase pressure in the transfer case

    Answer A is incorrect. Lubricant is added through the fill plug.

    Answer B is correct. The remote vent should be positioned at an elevated level to prevent water from entering the transfer case when fording water.

    Answer C is incorrect. The remote vent allows excessive pressure to exit the transfer case without allowing water in when fording deep water.

    Answer D is incorrect. The remote vent decreases pressure in the transaxle, not increases.

TASK C.14

36. When removing the transaxle differential from the vehicle, which of the following would be the LEAST LIKELY component to be removed?

A. Axle shaft

B. Transaxle

C. Lower control arms

D. Clutch assembly

Answer A is incorrect. The axle shaft must be removed anytime the differential is being removed from a transaxle.

Answer B is incorrect. The transaxle must be removed to remove the differential, because the transaxle is disassembled to remove the differential.

Answer C is incorrect. The lower control arms are removed or disconnected from the lower strut in order to remove the transaxle.

Answer D is correct. The clutch assembly does not need to be removed when removing the differential from a transaxle.

TASK E.2.3

37. When replacing a differential, Technician A says that the differential side bearings showing no damage can be reused with the new differential. Technician B says that if the differential side bearings are replaced, then the bearing races must be replaced. Who is correct?

A. A only

B. B only

C. Both A and B

D. Neither A nor B

Answer A is incorrect. Technician B is also correct.

Answer B is incorrect. Technician A is also correct.

Answer C is correct. Both technicians are correct. When replacing a differential, all bearings should be inspected and checked for smooth operation. If a bearing is found to be free from any defects, then it can be reused with the new differential. If the bearings show any questionable sign of wear, then they should be replaced, including new races.

Answer D is incorrect. Both technicians are correct.

TASK F.7

38. On a vehicle engaged in 4WD, all of the following conditions may cause a vibration that is more noticeable when changing throttle position EXCEPT:

A. Worn universal joints

B. Incorrect drive shaft angles

C. Tight drive shaft slip-joint splines

D. Worn front-drive axle joints

Answer A is incorrect. Worn universal joints will cause vibration that is more noticeable when changing throttle position.

Answer B is incorrect. Excessive drive shaft universal-joint working angles will cause vibration that might be more noticeable when changing throttle position.

Answer C is correct. A tight drive shaft slip-joint spline will have no effect on the driveline vibration. A seized-up slip joint, however, may cause vibration problems.

Answer D is incorrect. Worn front-drive axle universal joints will cause a vibration that is more noticeable when changing throttle position.

39. While inspecting a reverse idler gear from a manual transmission, technician A says that the center bore for the reverse idler should be checked for a smooth mar-free surface. Technician B says that the reverse idler gear is splined in the center bore and that the splines should be checked for excessive wear or damage. Who is correct?

TASK B.11

    A. A only
    B. B only
    C. Both A and B
    D. Neither A nor B

    Answer A is correct. Only technician A is correct. The reverse idler could use individual needle rollers or a needle-roller bearing cage. All center bores should be inspected for a smooth mar-free surface.

    Answer B is incorrect. The reverse idler is not splined to anything. It is free-rotating on a reverse idler shaft, with needle roller bearings.

    Answer C is incorrect. Only technician A is correct.

    Answer D is incorrect. Technician A is correct.

40. Technician A says low range in many transfer cases is available only while in four-wheel drive mode. Technician B says many transfer cases send the power flow through a planetary gear set when in low range. Who is correct?

TASK F.2

    A. A only
    B. B only
    C. Both A and B
    D. Neither A nor B

    Answer A is incorrect. Technician B is also correct.

    Answer B is incorrect. Technician A is also correct.

    Answer C is correct. Both technicians are correct. Low range cannot be engaged in most transfer cases without the transfer case being in four-wheel drive mode. Some transfer cases achieve low range by transferring the power flow through a planetary gear set.

    Answer D is incorrect. Both technicians are correct.

PREPARATION EXAM 5—ANSWER KEY

1.	B	21.	A
2.	C	22.	B
3.	C	23.	C
4.	A	24.	A
5.	A	25.	C
6.	C	26.	D
7.	C	27.	B
8.	B	28.	D
9.	C	29.	C
10.	C	30.	C
11.	C	31.	A
12.	B	32.	B
13.	A	33.	C
14.	B	34.	C
15.	A	35.	C
16.	D	36.	D
17.	D	37.	D
18.	A	38.	D
19.	C	39.	C
20.	A	40.	B

PREPARATION EXAM 5—EXPLANATIONS

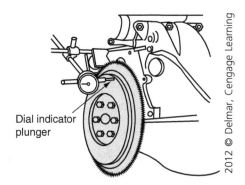

Dial indicator
plunger

2012 © Delmar, Cengage Learning

1. With the dial indicator positioned as shown in the illustration, the measurement being
    performed is:

    A. Crankshaft runout

    B. Crankshaft end-play

    C. Crankshaft warpage

    D. Crankshaft preload

    TASK A.9

    Answer A is incorrect. The dial indicator is not set up to measure crankshaft runout;
    however, flywheel runout could be checked with the dial indicator set up as shown.

    Answer B is correct. The crankshaft end-play is being measured in the illustration.

    Answer C is incorrect. The dial indicator is not set up to measure crankshaft warpage.
    Flywheel runout, however, could be checked with the dial indicator set up as shown.

    Answer D is incorrect. Crankshafts are not set up to use or have any preload; they have a
    specific end-play.

2. A manual transmission jumps out of fourth gear. Technician A says there may be excessive
    end-play between the fourth-speed gear and its matching synchronizer. Technician B says
    the detent springs on the fourth-gear shift rail may be weak. Who is correct?

    A. A only

    B. B only

    C. Both A and B

    D. Neither A nor B

    TASK B.1

    Answer A is incorrect. Technician B is also correct.

    Answer B is incorrect. Technician A is also correct.

    Answer C is correct. Both technicians are correct. A weak detent spring would allow the
    transmission to jump out of gear, especially on rough terrain. The synchronizer is used to
    match the speed of the gear to the speed of the main shaft for gear engagement. The
    synchronizer also helps hold the transmission in gear. Excessive end-play between the speed
    gear and synchronizer could allow the transmission to jump out of gear.

    Answer D is incorrect. Both technicians are correct.

TASK C.12

3. Technician A says that the speedometer drive gear does not have to be replaced when a new speedometer cable is replaced. Technician B says that the drive gears come in different teeth numbers and using the wrong one can cause the speedometer readings to change. Who is correct?

    A. A only
    B. B only
    C. Both A and B
    D. Neither A nor B

    Answer A is incorrect. Technician B is also correct.

    Answer B is incorrect. Technician A is also correct.

    Answer C is correct. Both technicians are correct. The speedometer drive gear is usually made of plastic or nylon and comes in various teeth sizes. The drive gear does not get replaced when replacing the speedometer cable unless it is damaged. If the wrong one is used, then the speedometer will read too fast or too slow.

    Answer D is incorrect. Both technicians are correct.

TASK D.2

4. A new universal joint has been installed in a vehicle. Technician A says the universal joint should be greased until grease is purged from all four caps. Technician B says that the universal joint is pre-lubricated and that drive shaft phasing can be affected by the amount of grease installed. Who is correct?

    A. A only
    B. B only
    C. Both A and B
    D. Neither A nor B

    Answer A is correct. Only technician A is correct. After installing a new universal joint, the joint should be greased until the grease is seen purging from all four bearing caps.

    Answer B is incorrect. Phasing is the proper aligning of yokes on a two-piece drive shaft. Grease has no effect on phasing.

    Answer C is incorrect. Only technician A is correct.

    Answer D is incorrect. Technician A is correct.

TASK E.1.5

5. When adjusting the pinion depth, technician A advises to replace the collapsible spacer. Technician B says that pinion depth can also be adjusted by installing a selective pinion bearing race. Who is correct?

    A. A only
    B. B only
    C. Both A and B
    D. Neither A nor B

    Answer A is correct. Only technician A is correct. A new collapsible spacer should always be used when the pinion nut is removed, as well as a new pinion nut.

    Answer B is incorrect. Pinion depth is typically adjusted using shims under the inner pinion bearing.

    Answer C is incorrect. Only technician A is correct.

    Answer D is incorrect. Technician A is correct.

6. A four-wheel drive vehicle has had its tires replaced because of premature tire wear for the second time. Technician A says the tire load capacity could be too small for the vehicle. Technician B says the tires may be inflated to the incorrect pressure. Who is correct?

TASK F.12

   A. A only
   B. B only
   C. Both A and B
   D. Neither A nor B

   Answer A is incorrect. Technician B is also correct.

   Answer B is incorrect. Technician A is also correct.

   Answer C is correct. Both technicians are correct. Tires used with too small of a load capacity can wear prematurely as well as running tires under- or overinflated.

   Answer D is incorrect. Both technicians are correct.

7. A vehicle with a manual transmission has been towed into the shop with a no-start complaint. The technician verifies the complaint and finds that the vehicle does not crank. Technician A advises to check the clutch safety switch for an open circuit. Technician B says the clutch switch could be out of adjustment. Who is correct?

TASK A.2

   A. A only
   B. B only
   C. Both A and B
   D. Neither A nor B

   Answer A is incorrect. Technician B is also correct.

   Answer B is incorrect. Technician A is also correct.

   Answer C is correct. Both technicians are correct. The clutch safety switch prevents starter engagement unless the clutch pedal is engaged. A defective or out-of-adjustment clutch switch could cause a starter engagement problem.

   Answer D is incorrect. Both technicians are correct.

8. When diagnosing a DTC for the output-shaft speed sensor, which of the following should NOT be done next after reading the code?

TASK B.15

   A. Check the service manual.
   B. Install a new sensor.
   C. Inspect the connector.
   D. Inspect the harness for continuity.

   Answer A is incorrect. The service manual is a good place to start after retrieving the trouble codes.

   Answer B is correct. A diagnostic trouble code only means there is a problem in the circuit. A new speed sensor should not be replaced until the complete circuit is checked.

   Answer C is incorrect. Many trouble codes are the result of a poor connection or a break in the wiring harness.

   Answer D is incorrect. Harnesses can get frayed from rubbing against other components or just get high resistance from corrosion.

TASK C.10

9.　A transaxle has transaxle fluid leaking from the output shaft seal on an all-wheel drive vehicle; the technician removes the seal for inspection and cannot see any problem with the seal. Technician A advises to check for a damaged yoke seal surface. Technician B says the vent could be plugged. Who is correct?

　　A.　A only
　　B.　B only
　　C.　Both A and B
　　D.　Neither A nor B

Answer A is incorrect. Technician B is also correct.

Answer B is incorrect. Technician A is also correct.

Answer C is correct. A seal leak can occur due to a bad seal, a damaged yoke surface or even a plugged vent, causing excessive pressures.

Answer D is incorrect. Both technicians are correct.

TASK D.6

10.　A rear-wheel drive vehicle is being diagnosed for a speed-related vibration. Technician A says most joint working angles should not exceed 3 degrees. Technician B says two joint working angles on a common shaft should cancel each other out within 1 degree. Who is correct?

　　A.　A only
　　B.　B only
　　C.　Both A and B
　　D.　Neither A nor B

Answer A is incorrect. Technician B is also correct.

Answer B is incorrect. Technician A is also correct.

Answer C is correct. Both technicians are correct. The universal joint working angles should be at least ½ degree, no more than 3 degrees unless specified by the manufacturer, and they should cancel each other out within 1 degree.

Answer D is incorrect. Both technicians are correct.

TASK E.2.3

11.　Differential side-bearing preload is being adjusted. Technician A says that some RWD differentials require the selection of the correct shim thickness to adjust the side-bearing preload. Technician B says that some RWD differentials use threaded adjusters to set side-bearing preload. Who is correct?

　　A.　A only
　　B.　B only
　　C.　Both A and B
　　D.　Neither A nor B

Answer A is incorrect. Technician B is also correct.

Answer B is incorrect. Technician A is also correct.

Answer C is correct. Both technicians are correct. On some differentials, side-bearing preload is determined by shim thickness behind the side-bearing race, and some differentials use threaded adjusters to set side-bearing preload.

Answer D is incorrect. Both technicians are correct.

12. A transfer case is being removed for servicing. Technician A says the transfer case and transmission must be removed as a unit. Technician B says all yokes and flanges should be marked for correct reassembly. Who is correct?

    TASK F.3

    A. A only
    B. B only
    C. Both A and B
    D. Neither A nor B

    Answer A is incorrect. Transfer cases and transmissions do not have to be removed as a unit.

    Answer B is correct. Only technician B is correct. All yokes and flanges should be marked for correct reassembly.

    Answer C is incorrect. Only technician B is correct.

    Answer D is incorrect. Technician B is correct.

13. When replacing a clutch, the pressure plate is found to have small cracks. Technician A says the pressure plate should be replaced. Technician B says the pressure plate should be resurfaced and then installed. Who is correct?

    TASK A.5

    A. A only
    B. B only
    C. Both A and B
    D. Neither A nor B

    Answer A is correct. Only technician A is correct. The pressure plate should be replaced with a new one.

    Answer B is incorrect. Unlike flywheels, pressure plates are not resurfaced; they are replaced.

    Answer C is incorrect. Only technician A is correct.

    Answer D is incorrect. Technician A is correct.

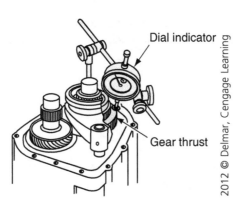

Dial indicator

Gear thrust

2012 © Delmar, Cengage Learning

TASK B.8

14. Technician A says transmission preload is being measured in the illustration. Technician B says if the measurement is excessive, then the thrust washer can be replaced to correct the measurement.

 A. A only
 B. B only
 C. Both A and B
 D. Neither A nor B

Answer A is incorrect. Transmission speed gears are not set with preload, and preload is not measured with a dial indicator.

Answer B is correct. Only technician B is correct. If the thrust on the speed gear is excessive, then a selective thrust washer or selective snap-ring can be installed to get the measurement in tolerance.

Answer C is incorrect. Only technician B is correct.

Answer D is incorrect. Technician B is correct.

TASK C.1

15. A four-speed transaxle has a clunking noise when driven in first gear and reverse gear. Technician A says that the gear on the countershaft could be the problem. Technician B says that the reverse idler gear could be the problem. Who is correct?

 A. A only
 B. B only
 C. Both A and B
 D. Neither A nor B

Answer A is correct. Only technician A is correct. A damaged cluster/countershaft gear could cause a clunking noise in first and reverse gears.

Answer B is incorrect. The reverse idler gear does not make contact in first gear; therefore, it would not make any noise in first gear.

Answer C is incorrect. Only technician A is correct.

Answer D is incorrect. Technician A is correct.

16. A clicking noise is heard on a front-wheel drive vehicle while turning a corner. The cause of this problem could be a bad

    A. Inner constant-velocity joint

    B. Front axle

    C. Wheel bearing

    D. Outer constant-velocity joint

TASK D.1

Answer A is incorrect. A bad inner constant-velocity joint would make a vibration or shudder on acceleration.

Answer B is incorrect. A bad front axle would make noise all of the time, not just on a turn.

Answer C is incorrect. A bad wheel bearing would cause a growling sound that got louder as the vehicle is steered left or right.

Answer D is correct. A bad outer constant-velocity joint will cause a clicking noise when turning.

17. A limited-slip differential is not working properly. Which of the following would be the LEAST LIKELY cause?

    A. Worn friction disc

    B. Weak preload spring

    C. Wrong oil additive used

    D. Incorrect shim used on the side bearing

TASK E.3.1

Answer A is incorrect. A worn friction disc would allow the limited-slip differential to slip too easily.

Answer B is incorrect. A weak preload spring would prevent the correct amount of preload to be placed on the side gears and allow them to slip too easily.

Answer C is incorrect. If the wrong additive is used in a limited-slip differential, then the friction discs could slip too easily.

Answer D is correct. A side-bearing preload problem would cause a noise, but would not affect limited-slip operation.

18. On a four-wheel drive vehicle with a straight front axle, front universal joint working angle can be corrected by

    A. Shimming the front axle housing

    B. Installing a shorter drive shaft

    C. The joint working angles cannot be changed

    D. Installing a 4-inch body lift

TASK F.8

Answer A is correct. Front U-joint working angles can be corrected with the use of shims under the front axle housing.

Answer B is incorrect. Front U-joint working angles cannot be corrected by installing a shorter drive shaft.

Answer C is incorrect. The front U-joint working angles can be changed.

Answer D is incorrect. Front U-joint working angles cannot be fixed by installing a 4-inch lift.

TASK A.1

19. A defective or worn pilot bearing may cause a rattling and growling noise under which condition?

    A. The engine is idling, with the clutch engaged.
    B. The engine is idling, and the clutch pedal is disengaged.
    C. The engine is idling, and the clutch is fully disengaged.
    D. The engine is idling, and the clutch pedal is fully disengaged.

    Answer A is incorrect. If the clutch is engaged and the engine idling, then the pilot bearing and transmission input shaft are rotating together.

    Answer B is incorrect. If the clutch pedal is disengaged, then the clutch is engaged, so both the pilot bearing and input shaft are rotating together.

    Answer C is correct. If the clutch is disengaged, the transmission input shaft and the pilot bearing are rotating at different speeds. After a short spin-down, the input shaft should not be spinning at all with the clutch fully disengaged. This would produce the most noise from a defective pilot bearing.

    Answer D is incorrect. If the clutch pedal is disengaged, then the clutch is engaged, so both the pilot bearing and input shaft are rotating together.

TASK B.2

20. Technician A says that many transmissions with internal linkage have no internal linkage adjustments. Technician B says that only transmissions with external linkage can be adjusted. Who is correct?

    A. A only
    B. B only
    C. Both A and B
    D. Neither A nor B

    Answer A is correct. Only technician A is correct. Most transmissions with internal linkage have no adjustments.

    Answer B is incorrect. While most transmissions with internal linkage have no linkage adjustment, this is not true for all transmissions.

    Answer C is incorrect. Only technician A is correct.

    Answer D is incorrect. Technician A is correct.

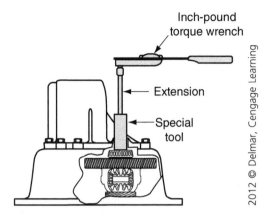

Inch-pound
torque wrench

Extension

Special
tool

2012 © Delmar, Cengage Learning

21. Technician A says that differential-bearing preload is being measured in the above
    illustration. Technician B says the final side-gear fastener torque is being applied. Who is
    correct?

    A. A only
    B. B only
    C. Both A and B
    D. Neither A nor B

TASK C.15

Answer A is correct. Only technician A is correct. An inch-pound torque wrench is typically
used to determine the rotating torque required (preload). This preload is adjusted with the
use of shims.

Answer B is incorrect. The side gears are not fastened with any type of fastener. Differential
bearing preload is being measured.

Answer C is incorrect. Only technician A is correct.

Answer D is incorrect. Technician A is correct.

22. When installing a replacement U-joint that has a grease zerk in the cross, the grease zerk
    should point toward the

    A. Yoke
    B. Drive shaft
    C. Engine
    D. Differential

TASK D.2

Answer A is incorrect. The grease zerk/fitting should be toward the drive shaft. If installed
away from the drive shaft, then the grease zerk/fitting may not be accessible.

Answer B is correct. It does not matter if the front or rear universal joint is being replaced;
the grease zerk/fitting goes toward the drive shaft.

Answer C is incorrect. It does not matter if the front or rear universal joint is being replaced;
the grease zerk/fitting goes toward the drive shaft.

Answer D is incorrect. It does not matter if the front or rear universal joint is being replaced;
the grease zerk/fitting goes toward the drive shaft.

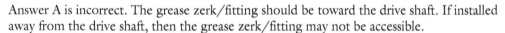

TASK E.4.3

23. Technician A says to rotate the axle slowly when checking the rear-axle runout. Technician B says to use a dial indicator to measure runout. Who is correct?

A.   A only

B.   B only

C.   Both A and B

D.   Neither A nor B

Answer A is incorrect. Technician B is also correct.

Answer B is incorrect. Technician A is also correct.

Answer C is correct. Both technicians are correct. The stylus of the dial indicator is positioned just outside the wheel studs on the axle flange. Rotate the axle slowly and note the dial indicator movement. This is the amount of axle flange runout. A runout of 0.005 inches (0.1mm) or less is acceptable.

Answer D is incorrect. Both technicians are correct.

TASK F.12

24. A vehicle with full-time, four-wheel drive has a different size tire on every wheel. Technician A says this could cause all three differentials to fail prematurely. Technician B says if the owner puts in special additive, then the differentials will be OK. Who is correct?

A.   A only

B.   B only

C.   Both A and B

D.   Neither A nor B

Answer A is correct. Only technician A is correct. Using different size tires at each wheel will result in all three differentials working constantly, which can lead to premature differential failure

Answer B is incorrect. There is no special additive to allow the use of different size tires.

Answer C is incorrect. Only technician A is correct.

Answer D is incorrect. Technician A is correct.

TASK A.1

25. Technician A says that clutch chatter could be caused by an uneven flywheel. Technician B says that oil on the clutch disc can cause clutch chatter. Who is correct?

A.   A only

B.   B only

C.   Both A and B

D.   Neither A nor B

Answer A is incorrect. Technician B is also correct.

Answer B is incorrect. Technician A is also correct.

Answer C is correct. Both technicians are correct. If the flywheel had an uneven surface, then it would not make full contact with the clutch disc, causing a possible chatter. Oil on the clutch disc would allow it to grab better at some spots than others, causing chatter.

Answer D is incorrect. Both technicians are correct.

26. A transmission will not engage in first gear when shifted. Which of these is the LEAST LIKELY cause?

    A. A broken synchronizer sleeve
    B. Misadjusted linkage
    C. A broken shift fork
    D. A broken tooth on the first gear

TASK B.1

Answer A is incorrect. A broken synchronizer sleeve would prevent gear engagement.

Answer B is incorrect. If the shift linkage was misadjusted, then the transmission may not go into gear.

Answer C is incorrect. If the shift fork were broken, then there would be no gear engagement.

Answer D is correct. A broken gear tooth would cause noise, but the transmission would still go into gear.

27. Technician A says that the input shaft end-play does not need to be checked before disassembling a transaxle. Technician B says to always rotate the input shaft to check turning effort before disassembling the transaxle. Who is correct?

    A. A only
    B. B only
    C. Both A and B
    D. Neither A nor B

TASK C.5

Answer A is incorrect. The input shaft end-play should always be checked before disassembly.

Answer B is correct. Only technician B is correct. Two checks should be made on a transaxle before disassembly. The input shaft end-play and the turning effort are needed to turn the input shaft.

Answer C is incorrect. Only technician B is correct.

Answer D is incorrect. Technician B is correct.

28. Technician A says that drive shaft runout should be checked with a digital caliper at the middle of the drive shaft. Technician B says that a dial indicator should be set at the differential end of the drive shaft to check runout. Who is correct?

    A. A only
    B. B only
    C. Both A and B
    D. Neither A nor B

TASK D.5

Answer A is incorrect. Digital calipers are not used to measure drive shaft runout. A dial indicator is used.

Answer B is incorrect. When measuring drive shaft runout, the drive shaft should be measured at the center and at each end about 3 inches from the ends.

Answer C is incorrect. Neither technician is correct.

Answer D is correct. Neither technician is correct. When measuring the drive shaft for runout, it should be measured using a dial indicator at the center and at both ends about 3 inches from the ends.

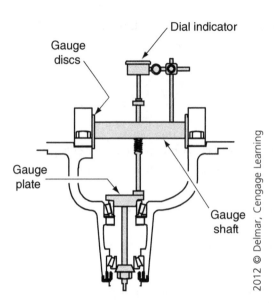

29. In the figure shown, after the dial indicator is rotated to the zero position with the stem on the gauge plate and then moved off the gauge plate, the dial indicator pointer moves 0.057 inches (1.45 mm) counterclockwise and the pinion gear is marked –4. The proper pinion depth shim is:

TASK E.1.5

A.   0.043 inch (1.09 mm)

B.   0.042 inch (1.07 mm)

C.   0.039 inch (0.99 mm)

D.   0.041 inch (1.04 mm)

Answer A is incorrect. A 0.043-inch (1.09 mm) shim would be too thick.

Answer B is incorrect. A 0.042-inch (1.07 mm) shim would be too thick.

Answer C is correct. The reading on the dial indicator must be subtracted from 0.100 inch to obtain the normal pinion-depth shim thickness. Therefore, the nominal shim thickness is 0.043 inch and the pinion marking of –4 is subtracted from this figure.

Answer D is incorrect. A 0.041-inch (1.04 mm) shim would be too thick.

30. When removing a transfer case from the vehicle, which of the following components listed is LEAST LIKELY to be disconnected or removed?

TASK F.3

A.   The front drive shaft

B.   The rear drive shaft

C.   The transmission

D.   The linkage

Answer A is incorrect. The front drive shaft must be disconnected from the transfer case to remove the case.

Answer B is incorrect. The rear drive shaft must be disconnected from the transfer case to remove the case.

Answer C is correct. Most transfer cases can be removed without removing the vehicle transmission.

Answer D is incorrect. Linkage must be disconnected from the transfer case.

31. Technician A says constant-running release bearings are used with hydraulically controlled clutches. Technician B says on a push-type pressure plate, the release bearing moves away from the pressure plate to disengage the clutch. Who is correct?

TASK A.4

   A.   A only
   B.   B only
   C.   Both A and B
   D.   Neither A nor B

   Answer A is correct. Only technician A is correct. Constant-running release bearings are used with hydraulically controlled clutches.

   Answer B is incorrect. The release bearings move toward the pressure plate to disengage the clutch.

   Answer C is incorrect. Only technician A is correct.

   Answer D is incorrect. Technician A is correct.

32. The clearance on the third-speed gear blocking ring is less than specified. Technician A says this may result in noise while driving in third gear. Technician B says this problem may cause hard shifting into third gear. Who is correct?

TASK B.9

   A.   A only
   B.   B only
   C.   Both A and B
   D.   Neither A nor B

   Answer A is incorrect. The blocker ring is part of a synchronizer assembly and blocks the shift until the speed of the speed gear and synchronizer hub are synchronized. It would not cause noise while driving in third gear.

   Answer B is correct. Only technician B is correct. If the blocker ring was worn below the allowed specification, then it would not be able to match the speed of the speed gear and synchronizer hub, causing a hard shift.

   Answer C is incorrect. Only technician B is correct.

   Answer D is incorrect. Technician B is correct.

33. Technician A says that most transaxle cases are sealed with room temperature vulcanizing (RTV) sealant. Technician B says that some transaxle cases have paper gaskets. Who is correct?

TASK C.3

   A.   A only
   B.   B only
   C.   Both A and B
   D.   Neither A nor B

   Answer A is incorrect. Technician B is also correct.

   Answer B is incorrect. Technician A is also correct.

   Answer C is correct. Both technicians are correct. RTV is used on many transaxles to seal the mating surfaces. RTV cures at room temperature. Anaerobic sealer can also be used. Anaerobic sealer cures in the absence of air. Some transaxles still use paper gaskets. It all depends on the manufacturer.

   Answer D is incorrect. Both technicians are correct.

TASK E.2.3

34. To measure differential case/carrier runout, which of the following tools would be used?

A. Micrometer

B. Feeler gauge

C. Dial indicator

D. Torque wrench

Answer A is incorrect. A micrometer is used to measure the thickness of a component.

Answer B is incorrect. A feeler gauge is used to measure the gap between two components.

Answer C is correct. The dial indicator is used to measure runout and end-play of components as well as backlash between two gears.

Answer D is incorrect. The torque wrench is used to torque fasteners and measure the preload on a bearing.

TASK F.10

35. A vehicle with four-wheel drive is in the shop for front-axle seal leakage. Upon draining the fluid, the technician notices that the fluid is milky and watery. Technician A says the condition could be from normal operation in high waters. Technician B says the vent location should be checked to make sure it has not broken and the hose is not split. Who is correct?

A. A only

B. B only

C. Both A and B

D. Neither A nor B

Answer A is incorrect. Technician B is also correct.

Answer B is incorrect. Technician A is also correct.

Answer C is correct. Both technicians are correct. Milky colored fluid is an indication of water mixed with the oil. Operating a four-wheel drive in high waters can cause this, but if the vent is broken or if the hose is cracked or split, then the condition will be aggravated.

Answer D is incorrect. Both technicians are correct.

TASK B.4

36. Technician A says that a transmission that is binding internally should be installed and allowed to break-in. Technician B says that if the transmission gets hung up during installation, then it should be drawn up the rest of the way using the bolts. Who is correct?

A. A only

B. B only

C. Both A and B

D. Neither A nor B

Answer A is incorrect. It is never OK to install a transmission that is binding. The problem needs to be found before installation. It is not normal for a transmission to bind after rebuild.

Answer B is incorrect. When installing a transmission, the transmission should be pushed until it is within a faction of an inch of being against the bell housing. Using the bolts to draw up a transmission can result in broken transmission cases or distorted bell housings.

Answer C is incorrect. Neither technician is correct.

Answer D is correct. Neither technician is correct. When rebuilding a transmission, be sure the transmission rotates freely before installation. During installation, do not use the attaching bolts to draw the transmission up into place.

37. Each of the following statements about manual transaxle shift linkages are true EXCEPT:

    A. Shift linkages may be cable operated.
    B. Shift linkage cables are longer on transaxles than on rear-wheel drive transmissions.
    C. Shift linkage may be rod-operated.
    D. All shift linkages are adjustable.

    TASK C.2

    Answer A is incorrect. Some transaxles use shift linkages that are cable-operated.

    Answer B is incorrect. The shift linkage cables on a transaxle are usually longer on a transaxle than on a rear-wheel drive transmission because the transaxle is further away from the shifting mechanism.

    Answer C is incorrect. Some transaxles use shift linkages that are rod-operated.

    Answer D is correct. Not all transaxle have adjustable shift linkage.

38. Technician A says the collapsible pinion-shaft spacer may be reused if the differential is disassembled and overhauled only once. Technician B says that after proper pinion bearing preload is set, pinion depth is adjusted by backing off the pinion flange nut one-quarter turn. Who is correct?

    TASK E.1.4

    A. A only
    B. B only
    C. Both A and B
    D. Neither A nor B

    Answer A is incorrect. It is never OK to reuse a collapsible pinion-shaft spacer. A collapsible spacer should be used only one time and then discarded if taken back apart.

    Answer B is incorrect. Pinion depth is adjusted using shims, not by backing off the pinion nut.

    Answer C is incorrect. Neither technician is correct.

    Answer D is correct. Neither technician is correct. When setting up a rear end, a new collapsible spacer should be used, but only after correct pinion depth is set using shims.

39. Technician A says low oil level in the transfer case is a sign of a leak. Technician B says overfilling a transfer case can cause an increase in fluid temperature. Who is correct?

    TASK F.6

    A. A only
    B. B only
    C. Both A and B
    D. Neither A nor B

    Answer A is incorrect. Technician B is also correct.

    Answer B is incorrect. Technician A is also correct.

    Answer C is correct. Both technicians are correct. A low fluid level is a sign of a fluid leak somewhere in the transfer case. It could be a seal, gasket, or even a drain plug. Special care should be taken when filling the transfer case not to overfill it. A high level of fluid will allow excessive aeration of the fluid, which will allow the fluid to overheat because of the lack of temperature transfer through the fluid due to the extra air bubbles.

    Answer D is incorrect. Both technicians are correct.

TASK F.11

40. Technician A says that all linkages on an electronic-shift transfer case are external. Technician B says that an open in the shift switch circuit will prevent the transfer case from engaging. Who is correct?

    A. A only
    B. B only
    C. Both A and B
    D. Neither A nor B

Answer A is incorrect. The shift motor drives the shift linkages that are housed inside the transfer case.

Answer B is correct. Only technician B is correct. An open in the shift switch circuit would not power or signal the transfer case motor to shift the unit into 4WD.

Answer C is incorrect. Only technician B is correct.

Answer D is incorrect. Technician B is correct.

PREPARATION EXAM 6—ANSWER KEY

1.	D	**21.**	A
2.	D	**22.**	A
3.	C	**23.**	B
4.	B	**24.**	C
5.	D	**25.**	A
6.	C	**26.**	C
7.	A	**27.**	A
8.	C	**28.**	B
9.	C	**29.**	A
10.	C	**30.**	C
11.	C	**31.**	C
12.	B	**32.**	A
13.	B	**33.**	C
14.	C	**34.**	D
15.	A	**35.**	C
16.	C	**36.**	C
17.	B	**37.**	C
18.	B	**38.**	C
19.	A	**39.**	A
20.	A	**40.**	A

PREPARATION EXAM 6—EXPLANATIONS

1. Which of the following is the LEAST LIKELY cause of clutch chatter?

TASK A.1

   A.   Weak clutch-disc torsional springs

   B.   A rear main seal leaking

   C.   A transmission input shaft seal leaking

   D.   Excessive input shaft end-play

Answer A is incorrect. The torsional springs in a clutch disc are to reduce clutch chatter.

Answer B is incorrect. A leaking rear main seal can contaminate the clutch disc with engine oil, causing a chatter.

Answer C is incorrect. A leaking input shaft seal would allow transmission oil to contaminate the clutch disc, causing chatter.

Answer D is correct. Excessive input shaft end-play could cause a noise or shifting problems.

TASK B.13

2. Uneven wear of the extension-housing bushing will most likely be caused by:

   A. Worn speedometer drive and drive gears
   B. Excessive transmission main shaft end-play
   C. A plugged transmission vent opening
   D. Uneven mating surfaces on the extension housing

   Answer A is incorrect. A worn speedometer drive and drive gears would not cause uneven bushing wear.

   Answer B is incorrect. Main shaft end-play does not usually cause bushing wear.

   Answer C is incorrect. A plugged transmission vent opening would cause the fluid to overheat and overflow.

   Answer D is correct. Uneven mating surfaces can offset the extension housing and cause uneven bushing wear.

TASK C.2

3. Technician A says an improper shift linkage adjustment may cause the transaxle to jump out of gear under a load. Technician B says a broken engine mount may cause the transaxle to jump out of gear. Who is correct?

   A. A only
   B. B only
   C. Both A and B
   D. Neither A nor B

   Answer A is incorrect. Technician B is also correct.

   Answer B is incorrect. Technician A is also correct.

   Answer C is correct. Both technicians are correct. Shift linkage that is not adjusted correctly can cause problems such as jumping out of gear under a load because there is not enough travel to fully engage the gear. A broken or loose engine mount can cause the transaxle to jump out of gear as well.

   Answer D is incorrect. Both technicians are correct.

TASK D.5

4. To measure drive shaft runout, the technician should do all of the following EXCEPT

   A. Place the vehicle in neutral.
   B. Use a micrometer to check the drive shaft for parallelism.
   C. Clean the drive shaft at the spot where the measurement will be taken.
   D. Jack up the vehicle and place on jack stands.

   Answer A is incorrect. The vehicle should be placed in neutral so the drive shaft can be rotated during the runout check.

   Answer B is correct. The drive shaft is not measured for parallelism using a micrometer. The runout check is performed using a dial indicator.

   Answer C is incorrect. The spot that is to be measured should be cleaned. This would be the center of the drive shaft and 3 inches from each end.

   Answer D is incorrect. The vehicle will need to be jacked up and placed on jack stands to allow the drive shaft to be rotated. If a lift is available, then this operation can be performed with the vehicle on the lift.

5. Which of the following would be the LEAST LIKELY cause for the brakes to be contaminated by rear-end oil?

    A. A tight bearing-retainer ring
    B. The worn oil-seal collar
    C. The axle shaft bearings are worn
    D. The axle bearing retainer plate distorted

TASK E.4.1

Answer A is incorrect. The bearing retainer ring is supposed to fit tight.

Answer B is incorrect. The oil seal collar provides the surface for the seal to ride on. If it were to wear, then oil can leak past the seal.

Answer C is incorrect. A worn axle bearing on a semi-floating axle would allow excessive load to be placed on the axle seal, causing it to leak oil.

Answer D is correct. A distorted axle bearing retainer plate would not cause it to leak oil. Bearing retainer plates get distorted from removal.

6. The most likely cause of a manual-shift transfer case not shifting into four-wheel drive (4WD) is which of the following?

    A. The transfer-case oil level is low.
    B. The rear drive shaft universal joints are bound up.
    C. The shift linkage is out of adjustment.
    D. The electronic shift motor is bad.

TASK F.2

Answer A is incorrect. A low transfer-case oil level would not keep the transfer case from shifting, but left unattended would eventually cause noise and transfer-case failure.

Answer B is incorrect. A binding universal joint would cause a vibration but would not keep the transfer case from engaging.

Answer C is correct. If the shift linkage is out of adjustment, then it can prevent the transfer case from going into four-wheel drive.

Answer D is incorrect. A manual-shifted transfer case would not have an electronic shift motor.

7. A bearing-type noise begins to come from the clutch and transmission area of a vehicle just as the clutch is almost completely disengaged. There is no noise when the clutch pedal is initially depressed. Technician A says that the clutch pilot bearing may be worn out. Technician B says that the release bearing is may be worn out. Who is correct?

TASK A.6

    A. A only
    B. B only
    C. Both A and B
    D. Neither A nor B

Answer A is correct. Only technician A is correct. When the clutch is almost completely disengaged, the engine and transmission are disconnected, allowing the pilot bearing to rotate. The transmission input shaft will be stationary. A pilot bearing will not make noise until the clutch is almost completely disengaged.

Answer B is incorrect. A worn release bearing will make noise as soon as the release bearing comes in contact with the pressure plate.

Answer C is incorrect. Only technician A is correct.

Answer D is incorrect. Technician A is correct.

TASK B.7

8. A manual transmission, when in neutral, has a growling noise with the engine idling and the clutch engaged. The noise disappears when the clutch pedal is depressed. This noise could be caused by what?

   A. Pilot bearing in the crankshaft

   B. Input shaft and pilot bearing contact area

   C. Input shaft bearing

   D. Mainshaft bearing

   Answer A is incorrect. If a pilot bearing is damaged, then it will make more noise when the clutch is fully disengaged.

   Answer B is incorrect. If the input shaft pilot bearing contact surface is damaged, then it will make more noise with the clutch disengaged.

   Answer C is correct. The input shaft is not turning with the clutch depressed, but is turning with the clutch engaged.

   Answer D is incorrect. A main shaft bearing failure would make noise no matter what position the clutch was in.

TASKS C.18 & 19

9. The needle bearings between the output shaft and the output shaft gears are scored and blue from overheating on a transaxle. Technician A says the transaxle may have been filled with the wrong lubricant at its last service. Technician B says the oil passage that supplies oil to those bearings may be restricted or plugged. Who is correct?

   A. A only

   B. B only

   C. Both A and B

   D. Neither A nor N

   Answer A is incorrect. Technician B is also correct.

   Answer B is incorrect. Technician A is also correct.

   Answer C is correct. Both technicians are correct. A restricted or plugged oil passage will starve the bearings of lubricating oil, causing overheating of the bearings and eventually complete failure. If the incorrect lubricant is used in a transaxle, then the lubricant will not flow through the system as designed, and the system will not perform as intended.

   Answer D is incorrect. Both technicians are correct.

TASK D.1

10. Technician A says that a clunking noise while cornering could be caused by a bad constant-velocity (CV) joint. Technician B says a CV joint could also make noise when the vehicle is traveling straight. Who is correct?

    A. A only

    B. B only

    C. Both A and B

    D. Neither A nor B

    Answer A is incorrect. Technician B is also correct.

    Answer B is incorrect. Technician A is also correct.

    Answer C is correct. Both technicians are correct. A faulty outer constant-velocity joint would most likely cause noise when turning a corner, while an inner constant-velocity joint will usually make noise going straight on acceleration.

    Answer D is incorrect. Both technicians are correct.

11. Technician A says that if the differential is overfilled, then it can cause axle seals to leak. Technician B says that worn axle bearings can cause the axle seals to leak. Who is correct?

    A. A only

    B. B only

    C. Both A and B

    D. Neither A nor B

TASK E.1.1

Answer A is incorrect. Technician B is also correct.

Answer B is incorrect. Technician A is also correct.

Answer C is correct. Both technicians are correct. Overfilling the differential with oil would flood the axle seals, causing a leak. Axle seals are not designed to hold off large amounts of oil. If the bearing were worn too much, then it would allow the axle shaft to apply too much pressure on the seal, causing seal leakage due to seal wear.

Answer D is incorrect. Both technicians are correct.

12. Mechanical transfer case linkage is being adjusted. Technician A says the transfer case lever should be placed in four-wheel drive high. Technician B says the transfer case levers are indexed using a pin or drill bit. Who is correct?

    A. A only

    B. B only

    C. Both A and B

    D. Neither A nor B

TASK F.2

Answer A is incorrect. The transfer case lever should be placed in the neutral position when adjusting the linkage.

Answer B is correct. Only technician B is correct. When adjusting mechanical transfer case linkage, the lever is placed in neutral, the adjusters are loosened, and then a pin or drill bit of a specified size is used to hold the transfer case lever indexed while the adjusters are tightened.

Answer C is incorrect. Only technician B is correct.

Answer D is incorrect. Technician B is correct.

13. When checking flywheel runout, which one of the following is LEAST LIKELY to affect runout?

    A. Crankshaft end-play

    B. Ring-gear wear

    C. Flywheel fastener torque

    D. Dial indicator stylus placement

TASK A.9

Answer A is incorrect. The crankshaft end-play has a tremendous effect on the flywheel runout.

Answer B is correct. Ring-gear wear will not affect the runout reading but would cause starter engagement problems.

Answer C is incorrect. If the flywheel is not torqued correctly, then the flywheel runout will be affected.

Answer D is incorrect. When checking flywheel runout, the placement of the stylus is crucial for accurate readings.

TASK B.3

14. Technician A says that when installing a cork gasket, no added sealant is required. Technician B says that rubber gaskets should be installed without any added sealant. Who is correct?

A. A only

B. B only

C. Both A and B

D. Neither A nor B

Answer A is incorrect. Technician B is also correct.

Answer B is incorrect. Technician A is also correct.

Answer C is correct. Both technicians are correct. Whether a cork or rubber gasket is being used, both types are designed to be used alone without additional sealant. Some technicians use a sealant to help hold the gasket in place while it is being installed.

Answer D is incorrect. Both technicians are correct.

TASK C.1

15. When draining the fluid from a manual transaxle, a gold-colored material is seen in the fluid. The most likely cause is a worn

A. Blocker ring

B. Speed gear

C. Reverse idler

D. Countershaft

Answer A is correct. A gold-colored material in the oil is usually the result of a worn brass blocker ring.

Answer B is incorrect. A speed gear is made from steel and would not cause the gold-colored material.

Answer C is incorrect. A reverse idler gear is made from steel and would not cause the gold-colored material.

Answer D is incorrect. A countershaft is made from steel and would not cause the gold-colored material.

TASK D.4

16. Technician A says that a strobe light and screw-type hose clamps can be used to balance a drive shaft in the vehicle. Technician B says some drive shafts have match marks for balance. Who is correct?

A. A only

B. B. only

C. Both A and B

D. Neither A nor B

Answer A is incorrect. Technician B is also correct.

Answer B is incorrect. Technician A is also correct.

Answer C is correct. Both technicians are correct. Using a strobe-type wheel balancer and two screw-type hose clamps, a drive shaft can be balanced. The strobe light is used to identify the heavy spot on a drive shaft; then hose clamps are installed with the screw opposite the heavy spot.

Answer D is incorrect. Both technicians are correct.

17. When draining a limited-slip differential, a paper-like material is found in the fluid. What is the most likely cause?

    A. A worn limited-slip synchronizer

    B. Worn limited-slip disc

    C. Damaged axle seal

    D. Damaged pinion seal

    TASK E.3.2

    Answer A is incorrect. The synchronizer is used in the transmission/transaxle.

    Answer B is correct. If the limited-slip clutch disc becomes damaged, then the paper-like friction material would be evident in the fluid.

    Answer C is incorrect. A damaged axle seal would cause a leak and possibly show signs of rubber in the fluid.

    Answer D is incorrect. A damaged pinion seal would cause a leak and possibly show signs of rubber in the fluid.

18. A four-wheel drive vehicle with front half shafts has a torn CV boot. What is the proper way to repair the boot?

    A. Replace the axle assembly.

    B. Replace the CV joint and boot.

    C. Repack the CV joint, and install a quick boot (split boot).

    D. Repack the CV joint, and reseal the boot with an approved sealer.

    TASK F.8

    Answer A is incorrect. Replacement of the entire axle shaft may not be necessary.

    Answer B is correct. The CV joint and the boot should be replaced.

    Answer C is incorrect. A quick boot is not the proper repair procedure.

    Answer D is incorrect. A new boot must be installed.

19. Technician A says that the freeplay at the clutch pedal should be measured before attempting to adjust the clutch. Technician B says if the clutch is slipping badly, then it can be corrected by adjusting the clutch. Who is correct?

    A. A only

    B. B only

    C. Both A and B

    D. Neither A nor B

    TASK A.2

    Answer A is correct. Only technician A is correct. As the clutch disc wears, the pedal freeplay will decrease. The clutch pedal freeplay should always be checked before attempting to adjust the clutch.

    Answer B is incorrect. Once a clutch is badly slipping, the disc is usually damaged. Adjusting the clutch may not correct the problem. Disc replacement is usually required, but not before determining the root cause of the failure.

    Answer C is incorrect. Only technician A is correct.

    Answer D is incorrect. Technician A is correct.

TASK B.12

20. Transmission end-play can be adjusted by all of the following EXCEPT

A. Different length bolts

B. Thrust washers

C. Thrust shims

D. Selective snap rings

Answer A is correct. Different length bolts are not used to adjust end-play; however, the length and position of bolts should be noted when disassembling a transmission.

Answer B is incorrect. Many transmissions use thrust washers to adjust the end-play. Thrust washers can be purchased in different thicknesses for end-play correction.

Answer C is incorrect. Thrust shims can be used in conjunction with thrust washers to adjust end-play.

Answer D is incorrect. Some transmissions use selective snap rings that come in different thicknesses to adjust end-play.

TASK C.2

21. All of the following about manual transaxle shift linkage is true EXCEPT

A. Many transmissions use steering-column-mounted gear shift levers.

B. Shift linkage can be cable-operated

C. Shift linkage cables are longer for a transaxle than those on a rear-wheel drive transmission.

D. Shift linkage can be internal.

Answer A is correct. While steering-column gear shift levers were common in the 60s and 70s, they are not found on current production vehicles.

Answer B is incorrect. The shift linkage can be cable operated.

Answer C is incorrect. The cables on a transaxle shift linkage are much longer than cables used on a rear-wheel drive transmission due to the location of the transaxle/transmission.

Answer D is incorrect. Transaxles use external linkages, while others use internal linkages.

TASK D.3

22. Which statement is true concerning a center-support bearing?

A. It is usually a sealed bearing and is maintenance-free.

B. It should be greased similar to that of a universal joint.

C. It is part of the drive shaft and cannot be replaced separately.

D. It is used on all rear-wheel drive vehicles.

Answer A is correct. The center-support bearing is usually sealed and cannot be greased as a universal joint would be. A good marine-grade waterproof grease should be packed along the edge of a center-support bearing during installation to help keep contaminants out.

Answer B is incorrect. The center-support bearing typically does not have a grease fitting.

Answer C is incorrect. The center-support bearing can be replaced when worn.

Answer D is incorrect. Center-support bearings are used on vehicles with a long wheel base to help with torsional vibration, but all rear-wheel drive vehicles do not use them.

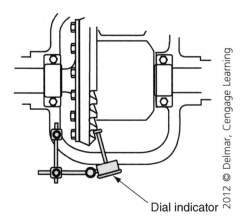

Dial indicator

23. In the figure shown, a technician is measuring:

A. Ring-gear runout
B. Ring-gear backlash
C. Bearing preload
D. Pinion gear backlash

TASK E.1.7

Answer A is incorrect. Ring-gear runout is not being checked. The dial indicator stylus would have to be on the back of the ring gear to measure runout.

Answer B is correct. Ring-gear backlash is measured off the heel of one of the ring-gear teeth.

Answer C is incorrect. Bearing preload is not being measured in the illustration.

Answer D is incorrect. Pinion gear backlash is not being measured.

24. Front universal joints are being replaced. Technician A says the new U-joints will still need to be greased after assembly. Technician B says if the universal joint is installed incorrectly, then the grease fitting cannot be reached with a grease gun. Who is correct?

A. A only
B. B only
C. Both A and B
D. Neither A nor B

TASK F.7

Answer A is incorrect. Technician B is also correct.

Answer B is incorrect. Technician A is also correct.

Answer C is correct. Both technicians are correct. After installing a new universal joint, the joint should be greased until grease is purged from all four caps. This insures that all needle bearings are completely lubricated. The grease fitting usually goes toward the shaft when installing universal joints.

Answer D is incorrect. Both technicians are correct.

TASK A.10

25. Technician A says sagged transmission mounts may cause improper drive shaft angles on a rear-wheel drive car. Technician B says improper drive shaft angles may cause a constant-speed vibration when the vehicle is accelerated and decelerated. Who is correct?

   A. A only
   B. B only
   C. Both A and B
   D. Neither A nor B

   Answer A is correct. Only technician A is correct. A sagging transmission mount would cause the universal joint working angles to change, possibly causing a vibration.

   Answer B is incorrect. A drive shaft vibration will not have a constant speed because the drive shaft speed is always changing with incorrect U-joint angles.

   Answer C is incorrect. Only technician A is correct.

   Answer D is incorrect. Technician A is correct.

TASK B.1

26. Excessive input shaft end-play in a 4-speed transmission may cause the transmission to jump out of:

   A. First gear
   B. Second gear
   C. Fourth gear
   D. Reverse gear

   Answer A is incorrect. First gear is at the rear of the transmission, away from the input shaft.

   Answer B is incorrect. Second gear is the second closest gear to the rear of the transmission.

   Answer C is correct. Fourth gear is the closest to the input shaft.

   Answer D is incorrect. Reverse gear is at the rear of the transmission.

TASK C.9

27. A transaxle shifts normally into all forward gears, but it will not shift into reverse gear; there is no evidence of noise while attempting this shift. Technician A says the reverse shifter fork may be broken. Technician B says the reverse idler gear teeth may be worn. Who is correct?

   A. A only
   B. B only
   C. Both A and B
   D. Neither A nor B

   Answer A is correct. Only technician A is correct. A broken shift fork would not usually cause noise.

   Answer B is incorrect. If the reverse-idler gear teeth were damaged, then there would be a noise.

   Answer C is incorrect. Only technician A is correct.

   Answer D is incorrect. Technician A is correct.

28. When replacing an axle boot, which component is LEAST LIKELY to be visually inspected?

    A. The constant-velocity joint
    B. A wheel bearing
    C. An axle shaft
    D. Axle seals

TASK D.2

Answer A is incorrect. The constant-velocity joint will be removed and cleaned to replace the boot and grease.

Answer B is correct. The wheel bearing is least likely to be inspected when replacing a constant-velocity boot.

Answer C is incorrect. While the constant-velocity joint is off the axle shaft, the spines should be inspected.

Answer D is incorrect. When the axle is out, the axle seals should be checked. If they are hard or show any damage, then they should be replaced.

29. When replacing the ring and pinion gears, all of the following must be replaced EXCEPT

    A. Spider gears
    B. Pinion seal
    C. Collapsible spacer
    D. Axle seals

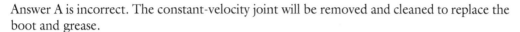

TASK E.1.4

Answer A is correct. The spider gears would not be replaced when replacing the ring and pinion.

Answer B is incorrect. The pinion seal should be replaced any time the pinion is removed.

Answer C is incorrect. A new collapsible spacer should be used when the pinion is removed.

Answer D is incorrect. Any time the axles are removed, the axle seals should be replaced.

30. A four-wheel drive vehicle has a jerking feeling in the steering wheel when turning sharply with the four-wheel drive engaged. Technician A says this could be normal for many four-wheel drive vehicles. Technician B says if the vehicle has CV joints in the front axle, then this would not be felt as much. Who is correct?

TASK F.7

    A. A only
    B. B only
    C. Both A and B
    D. Neither A nor B

Answer A is incorrect. Technician B is also correct.

Answer B is incorrect. Technician A is also correct.

Answer C is correct. Both technicians are correct. Many four-wheel drive vehicles use Cardan U-joints in the front outer axles; when turning sharply, they will cause a jerking feeling in the steering wheel due to the angle. CV joints can be turned as sharp as 40 degrees without the jerking feel that Cardan U-joints cause.

Answer D is incorrect. Both technicians are correct.

TASK A.2

31. Technician A says that a clutch may slip when it is out of adjustment. Technician B says that a transmission may grind when the clutch is out of adjustment. Who is correct?

    A. A only
    B. B only
    C. Both A and B
    D. Neither A nor B

    Answer A is incorrect. Technician B is also correct.

    Answer B is incorrect. Technician A is also correct.

    Answer C is correct. Both technicians are correct. Improper adjustment may not disengage the clutch, causing the gears to grind. Improper adjustment may not let the clutch fully engage.

    Answer D is incorrect. Both technicians are correct.

TASK B.1

32. Technician A says that a transmission that is hard to shift in gear may have a problem with the linkage not being lubricated. Technician B says that this problem may be caused by too strong a pressure plate installed in the vehicle's clutch system. Who is correct?

    A. A only
    B. B only
    C. Both A and B
    D. Neither A nor B

    Answer A is correct. Only technician A is correct. Dry or sticking linkage can cause hard shifting.

    Answer B is incorrect. Too strong a pressure plate would not cause hard shifting.

    Answer C is incorrect. Only technician A is correct.

    Answer D is incorrect. Technician A is correct.

TASK C.13

33. Technician A says that if side gears are damaged, then noise would only be heard when the vehicle is turned. Technician B says that side gears are shimmed for proper clearance. Who is correct?

    A. A only
    B. B only
    C. Both A and B
    D. Neither A nor B

    Answer A is incorrect. Technician B is also correct.

    Answer B is incorrect. Technician A is also correct.

    Answer C is correct. Both technicians are correct. The side gears only rotate when the vehicle is cornering. The side gears are shimmed for proper fit and clearance. The best way to check for wear is to measure clearances.

    Answer D is incorrect. Both technicians are correct.

34. Technician A says if the ring-gear tooth contact pattern is low at the toe of the tooth, then the pinion gear should be moved toward the ring gear. Technician B says if the pinion gear teeth have low flank contact on the ring-gear teeth, then the pinion gear should be moved toward the ring gear. Who is correct?

TASK E.1.6

   A. A only
   B. B only
   C. Both A and B
   D. Neither A nor B

Answer A is incorrect. The ring gear should be moved toward the pinion gear.

Answer B is incorrect. The drive pinion should be moved away from the ring gear.

Answer C is incorrect. Neither technician is correct.

Answer D is correct. Neither technician is correct. When the tooth contact pattern is high on the heel of the tooth or low on the toe of the tooth, the ring gear must be moved. If the tooth contact pattern is high or low on the tooth crest or root, then the pinion must be moved.

35. A vehicle with four-wheel drive is in the shop for front-axle seal leakage. Upon draining the fluid, the technician notices that the fluid is milky and watery. Technician A says the condition could be from normal operation in high waters. Technician B says the vent location should be checked to make sure it has not broken and that the hose is not split. Who is correct?

TASK F.10

   A. A only
   B. B only
   C. Both A and B
   D. Neither A nor B

Answer A is incorrect. Technician B is also correct.

Answer B is incorrect. Technician A is also correct.

Answer C is correct. Milky colored fluid is an indication of water mixed with the oil. Operating a four-wheel drive in high waters can cause this, but if the vent is broken or the hose cracked, then the condition will be aggravated.

Answer D is incorrect. Both technicians are correct.

36. A car with a 4-speed transmission has a growling noise coming from the transmission in all gears except fourth. In fourth, the noise is almost completely gone. Technician A says the input shaft bearing could cause the noise. Technician B says the countershaft bearing could be causing the noise. Who is correct?

TASK B.1

   A. A only
   B. B only
   C. Both A and B
   D. Neither A nor B

Answer A is incorrect. Technician B is also correct.

Answer B is incorrect. Technician A is also correct.

Answer C is correct. Both technicians are correct. The input shaft bearing would make noise in all gears EXCEPT fourth because in fourth gear the transmission is in direct drive. A bad countershaft bearing would make noise in all gears except direct drive (fourth).

Answer D is incorrect. Both technicians are correct.

TASK C.10

37. A transaxle has transaxle fluid leaking from the output shaft seal on an all-wheel drive vehicle; the technician removes the seal for inspection and cannot see any problem with the seal. Technician A says to check for a damaged yoke seal surface. Technician B says the vent could be plugged. Who is correct?

A. A only

B. B only

C. Both A and B

D. Neither A nor B

Answer A is incorrect. Technician B is also correct.

Answer B is incorrect. Technician A is also correct.

Answer C is correct. A seal leak can occur due to a bad seal, a damaged yoke surface, or even a plugged vent, causing excessive pressures.

Answer D is incorrect. Both technicians are correct.

TASK E.2.3

38. A vehicle with rear-wheel drive needs its spider gears replaced. Technician A says that some RWD axles require the removal of the pinion (spider) gear shaft in order to replace the spider gears. Technician B says that some RWD axles are held in place by bolts at the outer axle flange. Who is correct?

A. A only

B. B only

C. Both A and B

D. Neither A nor B

Answer A is incorrect. Technician B is also correct.

Answer B is incorrect. Technician A is also correct.

Answer C is correct. Some axles are held in place by large "C" clips that hold the axle in the side gear. Other axles are held in place by a flange and bolts at the backing plate.

Answer D is incorrect. Both technicians are correct.

TASK F.8

39. Technician A says an unbalanced drive shaft causes a vibration that increases as vehicle speed increases. Technician B says an unbalanced drive shaft causes a vibration that decreases as vehicle speed increases. Who is correct?

A. A only

B. B only

C. Both A and B

D. Neither A nor B

Answer A is correct. An unbalanced drive shaft causes a vibration that increases as vehicle speed increases.

Answer B is incorrect. An unbalanced drive shaft does not cause a vibration that decreases as vehicle speed increases.

Answer C is incorrect. Technician B is incorrect.

Answer D is incorrect. Technician A is correct.

40. A vehicle with all-wheel drive has a failed center differential. Technician A says mismatched tires could have caused this. Technician B says tires of the same size, but smaller than required, could have caused this. Who is correct?

TASK F.12

    A.   A only

    B.   B only

    C.   Both A and B

    D.   Neither A nor B

Answer A is correct. Mismatched tires can cause the center differential to fail due to constant operation.

Answer B is incorrect. Using four tires of a smaller but same size will not overwork the center differential.

Answer C is incorrect. Technician B is incorrect.

Answer D is incorrect. Technician A is correct.

PREPARATION EXAM ANSWER SHEET FORMS

ANSWER SHEET

1. _____	21. _____
2. _____	22. _____
3. _____	23. _____
4. _____	24. _____
5. _____	25. _____
6. _____	26. _____
7. _____	27. _____
8. _____	28. _____
9. _____	29. _____
10. _____	30. _____
11. _____	31. _____
12. _____	32. _____
13. _____	33. _____
14. _____	34. _____
15. _____	35. _____
16. _____	36. _____
17. _____	37. _____
18. _____	38. _____
19. _____	39. _____
20. _____	40. _____

ANSWER SHEET

1. _____   21. _____
2. _____   22. _____
3. _____   23. _____
4. _____   24. _____
5. _____   25. _____
6. _____   26. _____
7. _____   27. _____
8. _____   28. _____
9. _____   29. _____
10. _____   30. _____
11. _____   31. _____
12. _____   32. _____
13. _____   33. _____
14. _____   34. _____
15. _____   35. _____
16. _____   36. _____
17. _____   37. _____
18. _____   38. _____
19. _____   39. _____
20. _____   40. _____

ANSWER SHEET

1.	_____	21.	_____
2.	_____	22.	_____
3.	_____	23.	_____
4.	_____	24.	_____
5.	_____	25.	_____
6.	_____	26.	_____
7.	_____	27.	_____
8.	_____	28.	_____
9.	_____	29.	_____
10.	_____	30.	_____
11.	_____	31.	_____
12.	_____	32.	_____
13.	_____	33.	_____
14.	_____	34.	_____
15.	_____	35.	_____
16.	_____	36.	_____
17.	_____	37.	_____
18.	_____	38.	_____
19.	_____	39.	_____
20.	_____	40.	_____

ANSWER SHEET

1. _____   21. _____
2. _____   22. _____
3. _____   23. _____
4. _____   24. _____
5. _____   25. _____
6. _____   26. _____
7. _____   27. _____
8. _____   28. _____
9. _____   29. _____
10. _____   30. _____
11. _____   31. _____
12. _____   32. _____
13. _____   33. _____
14. _____   34. _____
15. _____   35. _____
16. _____   36. _____
17. _____   37. _____
18. _____   38. _____
19. _____   39. _____
20. _____   40. _____

ANSWER SHEET

1. _____   21. _____
2. _____   22. _____
3. _____   23. _____
4. _____   24. _____
5. _____   25. _____
6. _____   26. _____
7. _____   27. _____
8. _____   28. _____
9. _____   29. _____
10. _____   30. _____
11. _____   31. _____
12. _____   32. _____
13. _____   33. _____
14. _____   34. _____
15. _____   35. _____
16. _____   36. _____
17. _____   37. _____
18. _____   38. _____
19. _____   39. _____
20. _____   40. _____

ANSWER SHEET

1. _____   21. _____
2. _____   22. _____
3. _____   23. _____
4. _____   24. _____
5. _____   25. _____
6. _____   26. _____
7. _____   27. _____
8. _____   28. _____
9. _____   29. _____
10. _____   30. _____
11. _____   31. _____
12. _____   32. _____
13. _____   33. _____
14. _____   34. _____
15. _____   35. _____
16. _____   36. _____
17. _____   37. _____
18. _____   38. _____
19. _____   39. _____
20. _____   40. _____

Glossary

Actuator A device that delivers motion in response to an electrical signal.

A/D Converter Abbreviation for Analog-to-Digital Converter.

Additive An additive intended to improve a certain characteristic of the material or fluid.

Adsorber Catalyst An after-treatment technology that uses a base metal oxide and a precious metal compound as a catalyst to transform NOx to Nitrogen gas and H_2O (water vapor).

Aftercooler A charge air cooling device, usually water cooled.

Air Compressor An engine-driven mechanism for supplying high pressure air to the truck brake system.

Air Filter A device that minimizes the possibility of impurities entering the intake system.

Altitude Compensation System An altitude barometric switch and solenoid used to provide better drivability at 1,000 feet plus above sea level.

Ambient Temperature Temperature of the surrounding air. Normally, it is considered to be the temperature in the service area where testing is taking place.

Amp Acronym for ampere.

Ampere The unit for measuring electrical current.

Analog Signal A voltage signal that varies within a given range from high to low, including all points in between.

Analog-to-Digital Converter (A/D converter) A device that converts analog voltage signals to a digital format; located in the ECM.

Analog Volt/Ohmmeter (AVOM) A test meter used for checking voltage and resistance. Analog meters should not be used on solid state circuits.

Antifreeze A mixture added to water to lower its freezing point.

Armature The rotating component of a (1) starter or other motor; (2) generator.

Articulation Pivoting movement.

ASE Acronym for Automotive Service Excellence, a trademark of National Institute for Automotive Service Excellence.

Atmospheric Pressure The weight of the air at sea level; 14.696 pounds per square inch (psi) or 101.33 kilopascals (kPa).

Axis of Rotation The center line around which a gear or part revolves.

Battery Terminal A tapered post or threaded studs on top of the battery case for connecting the cables.

Bimetallic Two dissimilar metals joined together that have different bending characteristics when subjected to changes of temperature.

Blade Fuse A type of fuse having two flat male lugs for insertion into female box connectors.

Blower Fan A fan that pushes or blows air through a ventilation, heater, or air conditioning system.

Bobtailing A tractor running without a trailer.

Boss Outer race of a bearing.

Bottoming A condition that occurs when the teeth of one gear touch the lowest point between the teeth of a mating gear.

Brinelling A material surface failure caused by contact stress that exceeds the material limit. The result is a permanent dent or "brinell" mark.

British Thermal Unit (Btu) A measure of heat quantity equal to the amount of heat required to raise 1 pound of water 1° F.

Btu Acronym for British Thermal Unit.

CAA Acronym for Clean Air Act.

Cartridge Fuse A type of fuse having a strip of low melting point metal enclosed in a glass tube. If an excessive current flows through the circuit, the fuse element melts at the narrow portion, opening the circuit and preventing damage.

Caster The angle formed between the kingpin axis and a vertical axis as viewed from the side of the vehicle. Caster is considered positive when the top of the kingpin axis is behind the vertical axis.

Cavitation A condition caused by bubble collapse.

C-EGR Cooled exhaust gas recirculation.

CFC Acronym for chlorofluorocarbon.

Charging System A system consisting of the battery, alternator, voltage regulator, associated wiring, and the electrical loads of a vehicle. The purpose of the system is to recharge the battery whenever necessary and to provide the current required to power the electrical components.

Charging Circuit The alternator (or generator) and associated circuit used to keep the battery charged and power the vehicle's electrical system when the engine is running.

Check-Valve A valve that allows air to flow in one direction only.

Climbing A gear problem caused by excessive wear in gears, bearings, and shafts whereby the gears move sufficiently apart to cause the apex of the teeth on one gear to climb over the apex of another gear.

Clutch A device for connecting and disconnecting the engine from the transmission.

CO, or Carbon Monoxide One Carbon atom and one Oxygen atom. Produced on extremely small scale in diesel engines.

CO₂, or Carbon Dioxide One Carbon atom and two Oxygen atoms.

Coalescing Filter Media The filter media that removes the tiny aerosol particles from the blow by gas.

COE Acronym for cab-over-engine.

Coefficient of Friction A measurement of the amount of friction developed between two objects in physical contact when one of the objects is drawn across the other.

Coil Springs Spring steel spirals.

Common-Rail Fuel Delivery system that maintains a high-injection fuel-injection pressure regardless of engine speed, using high-pressure fuel stored in a single "common" rail or tube that connects to every fuel injector on the engine.

Compression Applying pressure to a spring or fluid.

Compressor Mechanical device that increases pressure within a circuit.

Condensation The process by which gas (or vapor) changes to a liquid.

Conductor Any material that permits the electrical current to flow.

Coolant Liquid that circulates in an engine cooling system.

Coolant Heater A component used to aid engine starting and reduce the wear caused by cold starting.

Coolant Hydrometer A tester designed to measure coolant specific gravity and determine antifreeze protection.

Cooling System System for circulating coolant.

Crankcase The housing within which the crankshaft rotates.

Cranking Circuit The starter circuit, including battery, relay (solenoid), ignition switch, neutral start switch (on vehicles with automatic transmission), and cables and wires.

Cross Head A device that allows a single rocker lever to depress two valves at a time.

Cycling (1) On-off action of the air conditioner compressor. (2) Repeated electrical cycling that can cause the positive plate material to break away from its grids and fall into the sediment base of the battery case.

Dampen To slow or reduce oscillations or movement.

Dampened Discs Discs that have dampening springs incorporated into the disc hub. When engine torque is transmitted to the disc, the plate rotates on the hub, compressing the springs. This action absorbs the torsional vibration caused by today's low RPM, high torque engines.

Data Links Circuits through which computers communicate with other electronic devices such as control panels, modules, sensors, or other computers.

Deburring To remove sharp edges from a cut.

Deflection Bending or moving to a new position as the result of an external force.

Detergent Additive An additive that helps keep metal surfaces clean and prevents deposits. These additives suspend particles of carbon and oxidized oil in the oil.

DER Acronym for Department of Environmental Resources.

Diagnostic Flow Chart A chart that provides a systematic approach to the electrical system and component troubleshooting and repair. They are found in service manuals and are vehicle make and model specific.

Dial Caliper A measuring instrument capable of taking inside, outside, depth, and step measurements.

Diesel Exhaust Fluid DEF is a urea-based chemical designed specifically for use in SCR systems to reduce NOx emissions. It is a mixture of urea and water.

Diesel Oxidation Catalyst A diesel oxidation catalyst (DOC) is a flow through device located in the exhaust system that consists of a canister containing a honeycomb-like structure. The substrate has a surface area that is coated with catalyst. As the exhaust gases goes through the catalyst, carbon monoxide and hydrocarbons are oxidized, thereby reducing harmful emissions.

Diesel Particulate Filter A diesel particulate filter, sometimes called a DPF, is a device mounted in the exhaust stream of a diesel engine. It is designed to remove diesel particulate matter or soot from the exhaust gas of a diesel engine.

Digital Binary Signal A signal that has only two values; on and off.

Digital Volt/Ohmmeter (DVOM) A test meter recommended for use on solid state circuits.

Diode Semiconductor device formed by joining P-type semiconductor material with N-type semiconductor material. A diode allows current to flow in one direction, but not in the opposite direction.

Dosing Injector The name of the DEF injector used to inject DEF in the exhaust stream of a diesel engine.

DOT Acronym for Department of Transportation.

Driven Gear A gear that is driven by a drive gear, by a shaft, or by some other device.

Drive, or Driving Gear A gear that drives another gear.

Driveline The propeller or driveshaft, and universal joints that links the transmission output to the axle pinion gear shaft.

Driveline Angle The alignment of the transmission output shaft, driveshaft, and rear axle pinion centerline.

Driveshaft Assembly of one or two universal joints connected to a shaft or tube; used to transmit torque from the transmission to the differential.

Drive Train An assembly that includes all torque transmitting components from the rear of the engine to the wheels.

ECM Acronym for electronic control module.

ECU Acronym for electronic control unit.

Eddy Current Circular current produced inside a metal core in the armature of a starter motor. Eddy currents produce heat and are reduced by using a laminated core.

Exhaust Gas Recirculation (EGR) Technology that diverts a small percentage of the exhaust gas back into the cylinder, lowering combustion temperatures and reducing NOx.

EGR Cooler A device which cools the EGR gases prior to entering the combustion chamber.

EGR Valve The valve which controls EGR flow.

Electricity The movement of electrons from one location to another.

Electromotive Force (EMF) The force that moves electrons between atoms. This force is the pressure that exists between the positive and negative points. This force is measured in units called volts. Charge differential.

Electronically Erasable Programmable Memory (EEPROM) Computer memory that enables write-to functions.

Electrons Negatively charged particles orbiting every nucleus.

EMF Acronym for electromotive force.

Engine Brake A hydraulically -operated device that converts the engine into a power absorbing mechanism.

Environmental Protection Agency An agency of the United States government charged with the responsibilities of protecting the environment.

EPA Acronym for the Environmental Protection Agency.

Exhaust After-Treatment Any technology which treats emissions in the exhaust flow, as opposed to inside the power cylinder.

Exhaust Brake A slide mechanism which restricts the exhaust flow, causing exhaust back pressure to build up in the engine's cylinders. The exhaust brake actually transforms the engine into a power absorbing air compressor driven by the wheels.

False Brinelling The polishing of a surface that is not damaged.

Fatigue Failures Progressive destruction of a shaft or gear teeth usually caused by overloading.

Fault Code A code that is recorded into the computer's memory.

Federal Motor Vehicle Safety Standard (FMVSS) A federal standard that specifies that all vehicles in the United States be assigned a Vehicle Identification Number (VIN).

Fixed Value Resistor An electrical device that is designed to have only one resistance rating, which should not change, for controlling voltage.

Flammable Any material that will easily catch fire or explode.

Flare To spread gradually outward in a bell shape.

Foot-Pound An English unit of measurement for torque. One foot-pound is the torque obtained by a force of 1 pound applied to a foot long wrench handle.

Fretting A result of vibration that the bearing outer race can pick up the machining pattern.

Fusible Link A term often used for an insulated fuse link.

Fuse Link A short length of smaller gauge wire installed in a conductor, usually close to the power source.

Gear A disk-like wheel with external or internal teeth that serves to transmit or change motion.

Gear Pitch The number of teeth per given unit of pitch diameter; an important factor in gear design and operation.

Ground The negatively charged side of a circuit. A ground can be a wire, the negative side of the battery, or the vehicle chassis.

Grounded Circuit A shorted circuit that causes current to return to the battery before it has reached its intended destination.

Harness and Harness Connectors The vehicle's electrical system wiring providing a convenient starting point for tracking and testing circuits.

Hazardous Materials Any substance that is flammable, explosive, or is known to produce adverse health effects in people or the environment.

Heads-Up Display (HUD) A technology used in some vehicles that superimposes data on the driver's normal field of vision. The operator can view the information, which appears to "float" just above the hood at a range near the front of a conventional tractor or truck. This allows the driver to monitor conditions such as road speed without interrupting his view of traffic.

Heater Control Valve A valve that controls the flow of coolant into the heater core from the engine.

Heat Exchanger A device used to transfer heat, such as a radiator or condenser.

Heavy-Duty Truck A truck that has a gross vehicle weight (GVW) of 26,001 pounds or more.

High-Resistant Circuits Circuits that have an increase in circuit resistance, with a corresponding decrease in current.

High-Strength Steel A low-alloy steel that is stronger than hot-rolled or cold-rolled sheet steels.

Hinged Pawl Switch The simplest type of switch; one that makes or breaks the current of a single conductor.

HUD Acronym for heads-up display.

Hydrometer A tester designed to measure the specific gravity of a liquid.

Inboard Toward the centerline of the vehicle.

In-Line Fuse A fuse that is in series with the circuit in a small plastic fuse holder, not in the fuse box or panel. It is used, when necessary, as a protection device for a portion of the circuit even though the entire circuit may be protected by a fuse in the fuse box or panel.

Installation Templates Drawings supplied by some vehicle manufacturers to allow the technician to correctly install an accessory. The templates available can be used to check clearances or to ease installation.

Insulator A material, such as rubber or glass, that offers high resistance to electron flow.

Integrated Circuit A component containing diodes, transistors, resistors, capacitors, and other electronic components mounted on a single piece of material and capable to perform numerous functions.

J1939 The standard used for communication and diagnostics among vehicle components.

Jacob's Engine Brake An engine brake named for its inventor. A hydraulically-operated device that converts a power producing diesel engine into a power-absorbing retarder.

Jumper Wire A wire used to temporarily bypass a circuit or components for electrical testing. A jumper wire consists of a length of wire with an alligator clip at each end.

Jump Start The procedure used when it becomes necessary to use a boost battery to start a vehicle with a discharged battery.

Kinetic Energy Energy in motion.

Lateral Runout The wobble or side-to-side movement of a rotating wheel.

Lazer Beam Alignment System A two- or four-wheel alignment system using wheel-mounted instruments to project a lazer beam to measure toe, caster, and camber.

Linkage A system of rods and levers used to transmit motion or force.

Low-Maintenance Battery A conventionally vented, lead/acid battery, requiring normal periodic maintenance.

Magnetorque An electromagnetic clutch.

Maintenance-Free Battery A battery that does not require the addition of water during normal service life.

Maintenance Manual A publication containing routine maintenance procedures and intervals for vehicle components and systems.

National Automotive Technicians Education Foundation (NATEF) A foundation having a program of certifying secondary and post-secondary automotive and heavy-duty truck training programs.

NATEF Acronym for National Automotive Education Foundation.

National Institute for Automotive Service Excellence (ASE) A nonprofit organization that has an established certification program for automotive, heavy-duty truck, auto body repair, engine machine shop technicians, and parts specialists.

NIOSH Acronym for National Institute for Occupation Safety and Health.

NLGI Acronym for National Lubricating Grease Institute.

NHTSA Acronym for National Highway Traffic Safety Administration.

NO, or Oxide of Nitrogen One nitrogen atom and one oxygen atom.

NO$_2$, or Nitrous Dioxide One nitrogen atom and two oxygen atoms. The 2 stands for the number of oxygen atoms.

NOx, or Oxides of Nitrogen One nitrogen atom and three or more oxygen atoms. The x stands for the unknown number of oxygen atoms.

NOP Acronym for nozzle opening pressure. Pressure in an injector nozzle opens at inoperation. Also known as VOP.

OEM Acronym for original equipment manufacturer.

Off-Road With reference to unpaved, rough, or ungraded terrain on which a vehicle will operate. Any terrain not considered part of the highway system falls into this category.

Ohm A unit of electrical resistance.

Ohm's Law Basic law of electricity stating that in any electrical circuit, current, resistance, and pressure work together in a mathematical relationship.

On-Road With reference to paved or smooth-graded surface on which a vehicle will operate; part of the public highway system.

Open Circuit An electrical circuit whose path has been interrupted or broken either accidentally (a broken wire) or intentionally (a switch turned off).

Oscillation Movement in either fore/aft or side-to-side direction about a pivot point.

OSHA Acronym for Occupational Safety and Health Administration.

Out of Round Eccentric.

Output Driver Electronic switch that the computer uses to control the output circuit. Output drivers are located in the output ECM.

Oval Condition that occurs when a tube is egg-shaped.

Overrunning Clutch A clutch mechanism that transmits power in one direction only.

Overspeed Governor A governor that shuts off fuel at a specific RPM.

Oxidation Inhibitor An additive used with lubricating oils to keep oil from oxidizing at high temperatures.

Oxide of Nitrogen, or NO One nitrogen atom and one oxygen atom.

Oxides of Nitrogen, or NOx One nitrogen atom and three or more oxygen atoms. The x stands for the unknown number of oxygen atoms.

Parallel Circuit An electrical circuit that provides two or more paths for current flow.

Parallel Joint Type A type of driveshaft installation whereby all companion flanges and/or yokes in the complete driveline are parallel to each other with the working angles of the joints of a given shaft being equal and opposite.

Parking Brake A mechanically applied brake used to prevent a parked vehicle's movement.

Particulate Matter The fine particles in diesel exhaust.

Parts Requisition A form that is used to order new parts on which the technician writes the part(s) needed along with the vehicle's VIN.

Payload The weight of the cargo carried by a truck, not including the weight of the body.

Pitting Surface irregularities resulting from corrosion.

Polarity The state, either positive or negative, of charge differential.

Pole The number of input circuits made by an electrical switch.

Pounds per Square Inch (psi) A unit of English measure for pressure.

Power A measure of work being done factored with time.

Power Flow The flow of power from the input shaft through one or more sets of gears.

Power Train Engine to the wheels in a vehicle.

Pressure The force applied to a definite area measured in pounds per square inch (psi) English, or kilopascals (kPa) metric.

Pressure Differential The difference in pressure between any two points of a system or a component.

Printed Circuit Board Electronic circuit board made of thin nonconductive material onto which conductive metal has been deposited. The metal is then etched by acid, leaving lines that form conductors for the circuits on the board. A printed circuit board can hold many complex circuits.

Programmable Read Only Memory (PROM) Electronic component that contains program information specific to vehicle model calibrations.

PROM Acronym for Programmable Read Only Memory.

psi Acronym for pounds per square inch.

P-Type Semiconductors Positively biased semiconductors.

RAM Acronym for random access memory; main memory.

Ram Air Air forced into the engine housing or passenger compartment by the forward motion of the vehicle.

Random Access Memory (RAM) Memory used during computer operation to store temporary information. The microcomputer can write, read, and erase information from RAM; electronically retained.

RCRA Acronym for Resource Conservation and Recovery Act.

Reactivity The characteristic of a material that enables it to react violently with air, heat, water, or other materials.

Read Only Memory (ROM) Memory used in microcomputers to store information permanently.

Recall Bulletin A bulletin that pertains to special situations that involve service work or replacement of components in connection with a recall notice.

Reference Voltage The voltage supplied to a sensor by the computer, which acts as base line voltage; modified by the sensor to act as an input signal, usually 5 VDC.

Relay An electric switch that allows a small current to control a much larger one. It consists of a control circuit and a power circuit.

Reserve Capacity Rating The ability of a battery to sustain a minimum vehicle electrical load in the event of a charging system failure.

Resistance Opposition to current flow in an electrical circuit.

Resource Conservation and Recovery Act (RCRA) Law that states that after using hazardous material, it must be properly stored until an approved hazardous waste hauler arrives to take it to a disposal site.

Revolutions per Minute (RPM) The number of complete turns a shaft turns in one minute.

Right to Know Law A law passed by the federal government and administered by the Occupational Safety and Health Administration (OSHA) that requires any company that uses or produces hazardous chemicals or substances to inform its employees, customers, and vendors of any potential hazards that may exist in the workplace as a result of using the products.

Ring Gear The gear around the edge of a flywheel.

ROM Acronym for read only memory.

Rotary Oil Flow A condition caused by the centrifugal force applied to the fluid as the converter rotates around its axis.

Rotation A term used to describe a gear, shaft, or other device when it is turning.

RPM Acronym for revolutions per minute.

Rotor Rotating part of the alternator that provides the magnetic fields necessary to create a current flow. The rotating member of an assembly.

Runout Deviation or wobble of a shaft or wheel as it rotates. Measured with a dial indicator.

Selective Catalyst Reduction Selective catalytic reduction (SCR) injects a small amount of urea into the diesel exhaust stream to convert Nox into N_2 and O_2.

Semiconductor Solid state material used in diodes and transistors.

Sensing Voltage The voltage that allows the regulator to sense and monitor the battery voltage level.

Sensor An electronic device used to monitor conditions for computer control requirements. An input circuit device.

Series Circuit A circuit connected to a voltage source with only one path for electron flow.

Series/Parallel Circuit Circuit designed so that both series and parallel conditions exist within the same circuit.

Service Bulletin Publication that provides the latest service tips, field repairs, product improvements, and related information of benefit to service personnel.

Service Manual A manual, published by the manufacturer, that contains service and repair information for all vehicle systems and components.

Short Circuit An undesirable connection between two worn or damaged wires. The short occurs when the insulation is worn between two adjacent wires and the metal in each wire contacts the other, or when wires are damaged or pinched.

Single-Axle Suspension A suspension with one axle.

Single Reduction Axle Any axle assembly that employs only one gear reduction through its differential carrier assembly.

Smart Device A device connected to, and able to communicate on, the J1939 data bus.

Solenoid An electromagnet that is used to conduct electrical energy in mechanical movement.

Solid Wires Single-strand conductors.

Solvent Substance that dissolves other substances.

Soot A general term that refers to impure carbon particles resulting from the incomplete combustion of a hydrocarbon.

Spade Fuse Term used for blade fuse.

Spalling Surface fatigue that occurs when chips, scales, or flakes of metal break off.

Specialty Service Shop A shop that specializes in areas such as engine rebuilding, transmission/axle overhauling, brake, air conditioning/heating repairs, and electrical/electronic work.

Specific Gravity The ratio of a liquid's mass to that of an equal volume of distilled water.

Spontaneous Combustion A reaction in which a combustible material self-ignites.

Stall Test Test performed when there is a malfunction in the vehicle's power package (engine and transmission), to determine which of the components is at fault.

Starter Circuit The circuit that carries the high current flow and supplies power for engine cranking.

Starter Motor Device that converts electrical energy from the battery into mechanical energy for cranking.

Starter Safety Switch Switch that prevents vehicles with automatic transmissions from being started in gear.

Static Balance Balance at rest, or still balance.

Stepped Resistor A resistor designed to have two or more fixed values, available by connecting wires to one of several taps.

Storage Battery A battery to provide a source of direct current electricity for both the electrical and electronic systems.

Stranded Wire Wire that is are made up of a number of small solid wires, generally twisted together, to form a single conductor.

Sulfation Condition that occurs when sulfate is allowed to remain on the battery plates for a long time, causing two problems: (1) It lowers the specific gravity levels, increasing the danger of freezing at low temperatures. (2) In cold weather, a sulfated battery may not have the reserve power needed to crank the engine.

Sulfur A natural element that has been linked to acid formation both inside engines and in the atmosphere.

Swage To reduce or taper.

Switch Device used to control on/off and direct the flow of current in a circuit. A switch can be under the control of the driver or can be self-operating through a condition of the circuit, the vehicle, or the environment.

Tachometer Instrument that indicates shaft rotating speeds.

Throw (1) Offset of a crankshaft. (2) Number of output circuits of a switch.

Time Guide Used for computing compensation payable by the truck manufacturer for repairs or service work to vehicles under warranty.

Timing The phasing of events to produce action such as ignition.

Torque Twisting force.

Torque Converter A device, similar to a fluid coupling, that transfers engine torque to the transmission input shaft and can multiply engine torque.

Toxicity A statement of how poisonous a substance is.

Tractor A motor vehicle that has a fifth wheel and is used for pulling a semitrailer.

Transistor Electronic device produced by joining three sections of semiconductor materials. Used as a switching device.

Tree Diagnosis Chart Chart used to provide a logical sequence for what should be inspected or tested when troubleshooting a repair problem.

Ultra-Low Sulfur Fuel (ULSF) Diesel fuel which contains less than 15 parts per million by volume of sulfur. Mandated phase-in starting in mid-2006.

Urea Used in SCR reactions to reduce the NOx pollutants in exhaust gases from combustion. Urea is carried on board the vehicle as a water solution in a storage tank.

Urea Injection and Control System A system and controls (including NOx and urea quality sensors) required to deliver a precise amount of urea under all environmental conditions.

Vacuum Pressure values below atmospheric pressure.

Vehicle Retarder An engine or driveline brake.

VG Turbo Variable Geometry Turbocharger Turbochargers that constantly adjust the amount of airflow into the combustion chamber, optimizing performance and efficiency.

VGT Actuator The computer-controlled device which opens and closes the variable geometry turbocharger.

VIN Acronym for vehicle identification number.

Viscosity Resistance to flow or fluid sheer.

VOP Acronym for Valve Opening Pressure. Caterpiller term for NOP.

Volt The unit of electromotive force.

Voltage-Generating Sensors Devices that produce their own voltage signal.

Voltage Limiter Device that provides protection by limiting voltage to the instrument panel gauges to approximately 5 volts.

Voltage Regulator Device that controls the current produced by the alternator and thus the voltage level in the charging circuit.

Watt Measure of electrical power.

Watt's Law A law of electricity used to calculate the power consumed in electrical circuit, expressed in watts. It states that power equals voltage multiplied by current.

Windings (1) The three bundles of wires in the stator. (2) Coil of wire in a relay or similar device.

Work (1) Forcing a current through a resistance. (2) The product of a force.

Yield Strength The highest stress a material can stand without permanent deformation or damage, expressed in pounds per square inch (psi).

Notes

Notes

Notes

Notes

Notes

Notes

Notes

Printed in the USA
CPSIA information can be obtained
at www.ICGtesting.com
JSHW062016270923
49159JS00011B/201